I0095131

Women and Tourist Work in Jamaica

Anthropology of Tourism: Heritage, Mobility, and Society

Series Editor: Michael A. Di Giovine
(West Chester University of Pennsylvania)

Mission Statement

The Anthropology of Tourism: Heritage, Mobility, and Society series provides anthropologists and others in the social sciences and humanities with cutting-edge and engaging research on the culture(s) of tourism. This series embraces anthropology's holistic and comprehensive approach to scholarship, and is sensitive to the complex diversity of human expression. Books in this series particularly examine tourism's relationship with cultural heritage and mobility and its impact on society. Contributions are transdisciplinary in nature, and either look at a particular country, region, or population, or take a more global approach. Including monographs and edited collections, this series is a valuable resource to scholars and students alike who are interested in the various manifestations of tourism and its role as the world's largest and fastest-growing source of socio-cultural and economic activity.

Advisory Board Members

Quetzil Castañeda, Saskia Cousin, Jackie Feldman, Nelson H. H. Graburn, Jafar Jafari, Tom Selwyn, Valene Smith, Amanda Stronza, Hazel Tucker, and Shinji Yamashita

Recent Titles in the Series

Women and Tourist Work in Jamaica: Seven Miles of Sandy Beach by A. Lynn Bolles
Encounters across Difference: Tourism and Overcoming Subalternity in India by Natalia Bloch
Study Abroad and the Quest for an Anti-Tourism Experience edited by John Bodinger de Uriarte and Michael A. Di Giovine
Tourism and Language in Vieques: An Ethnography of the Post-Navy Period by Luis Galanes Valldejuli
The Ethnography of Tourism: Edward Bruner and Beyond edited by Naomi Leite, Quetzil E. Castañeda, and Kathleen M. Adams
Capoeira, Mobility, and Tourism: Preserving an Afro-Brazilian Tradition in a Globalized World by Sergio González Varela

Women and Tourist Work in Jamaica

Seven Miles of Sandy Beach

A. Lynn Bolles

LEXINGTON BOOKS
Lanham • Boulder • New York • London

Published by Lexington Books
An imprint of The Rowman & Littlefield Publishing Group, Inc.
4501 Forbes Boulevard, Suite 200, Lanham, Maryland 20706
www.rowman.com

86-90 Paul Street, London EC2A 4NE

Copyright © 2022 by The Rowman & Littlefield Publishing Group, Inc.

All rights reserved. No part of this book may be reproduced in any form or by any electronic or mechanical means, including information storage and retrieval systems, without written permission from the publisher, except by a reviewer who may quote passages in a review.

British Library Cataloguing in Publication Information Available

Library of Congress Cataloging-in-Publication Data

Names: Bolles, Augusta Lynn, author.
Title: Women and tourist work in Jamaica : seven miles of sandy beach / A. Lynn Bolles.
Description: Lanham : Lexington Books, [2022] | Includes bibliographical references and index.
Identifiers: LCCN 2021041636 (print) | LCCN 2021041637 (ebook) | ISBN 9781793615565 (cloth) | ISBN 9781793615589 (paper) | ISBN 9781793615572 (ebook)
Subjects: LCSH: Women in tourism—Jamaica—Negril. | Tourism—Employees. | Women employees—Jamaica—Negril—Social conditions. | Tourism—Jamaica—Negril.
Classification: LCC G155.J25 B65 2021 (print) | LCC G155.J25 (ebook) | DDC 338.4/7917292—dc23/eng/20211008
LC record available at https://lccn.loc.gov/2021041636
LC ebook record available at https://lccn.loc.gov/2021041637

Contents

Preface

When people ask what I do for a living, I explain that I am a sociocultural anthropologist whose area of expertise is the Caribbean and that the people I work with are mostly women. The usual reply is in reference to the geographic region with comments such as "Oh what fun, what a great place!" "Why, we vacationed in Jamaica and Puerto Rico," and "Is Cancun in the Caribbean?" Further on when I mention that my fieldwork site is on the western tip of the island in Negril, the inquirers' reply rises in volume and enthusiasm by remarking, "Now you are talking! This is academic research, oh yeah, right?!"

Those who know something about ethnographic research understand that yes, it can be very rewarding in terms of theory and methodology, and can be professionally challenging and at times fun. In the end, ethnographic work is based on human encounters and interactions with other humans, their environment, and other aspects of their lives. It is extraordinarily in-your-face kind of hard work that crosses class, race/color/nationality, gender, and other differences on both sides of that human encounter. Hard work and contended forces are most salient in a place like Negril where one meets up with the community's women, whose livelihoods are predicated on their access to the United States, CAN dollar, the EURO, or whatever foreign currency is being exchanged. Together these human activities drive the tourist industry. The question then is how did I enter the anthropology of tourism arena and specifically in Jamaica? Of course, it began with a vacation to Jamaica.

A few years prior to my entering the field of anthropology as a graduate student, I took a long weekend trip with my mother to Montego Bay, Jamaica. Our off-season affordable trip accommodations put us in a hotel in need of refurbishing. A decade later, this property would launch the Sandals Resorts empire.

A return trip to Jamaica and my first trip to Negril occurred in 1977 while I was an anthropology graduate student member of a summer study abroad program (SUNY-Brockport). Organized by Jamaican-born historian Dr. Ena Farley and her husband, the esteemed Guyanese economist Dr. Rawle Farley, the group heard outstanding lectures by leading University of the West Indies scholars and took field trips across the island. On one of those trips, we went to Negril, located on the western tip of the island. At that time, Negril was a small, laid-back place that attracted international backpackers and Jamaican families on holiday who occupied the few number of small hotels and family-owned cottages to rent. Facing the decline in tourism due to the Cold War global politics of the day, the Government of Jamaica owned and operated an "adult singles only" all-inclusive package (prepaid access to food, beverages, and activities on property) resort. Like the rest of the country, the resort was also in a cash-flow crisis, so it welcomed one-time-only visitors such as us students. As a rookie participant observer, the resort was an eye-opener as to how the tourist industry successfully marketed adult pleasure of sea, sand, alcohol, sun, and skin although not necessarily in that order. Negril, once upon a time a fishing village, was now flirting with the world of global tourism.

A year later, while I was carrying out my dissertation research in Kingston, I took advantage of being a "local resident" tourist, as my temporary Jamaican resident visa afforded me a local discount package. Always the ethnographer, I participated being a tourist in Negril while observing the social and cultural environment. The village was slowly entering a phase of economic development fueled by the increase in numbers of luxury hotels and paved roads. However, the range of accommodations and experiences still kept with its origins as a fishing village on the outskirts of sugar plantations. I continued to be a tourist/participant/observer/adopted "Jamaican" by long association, returning to Negril now and then with my husband and growing family.

Fast forward to the late 1980s. My scholarly interest in tourism emerged from my research for *In the Shadows of the Sun* (Deere et al. 1990), a volume that examined the outcomes of structural adjustment policies (SAPs) on the lives of women and families and the national economies of the Caribbean. At that time, few academic studies had examined tourism in Jamaica, and most information was generated by the industry itself. Furthermore, the industry and the government were primarily interested in how to decrease the low-value dimension in tourist revenues. My interest in tourism began by assessing the gaps I found in the social science academic literature at hand, particularly in terms of women workers in the industry. I used my continued commitment to Jamaican feminist social science knowledge production that began almost twenty years earlier in the Kingston Metropolitan Area. Instead of the precarity of Jamaica's manufacturing sector of the 1970s and structural adjustment policies, this time the work site was on the beach, in lodgings and restaurants constrained by weather (hurricane season), successful public relations campaigns, and good word-of-mouth endorsements of returned visitors and travel agents. I would find out that global politics were as much of a factor in determining success in a country's tourism as were other international policy benefactors.

I am fortunate to be affiliated with since its founding the unit in the mid-1980s now known as the Institute of Gender and Development Studies (IGDS) at the University of the West Indies, Mona. IGDS is a degree-granting program focused on scholarship and research concerning issues of gender as it impacts society. The research for *Women and Tourist Work* and my other scholarly efforts have a home base at IGDS and its counterpart in Barbados, the Dame Nita Barrow IGDS.

Over the years, there were a variety of ways to keep abreast of what was happening in Negril. When I could not make my annual sojourn to Negril, telephone calls and letters to a few of the women with whom I worked allowed me to stay connected. However, this was not the case for the majority, whose location was fixed by their work site, which was most unreliable. Over time, due to the brevity of our encounters and conversations plus the passing of time, I lost contact, but their experiences are alive in my notes and writings. Once in a while when I heard of a colleague's plan to vacation in Negril, I cajoled them to carry messages from me to my friends there. On their return, my colleagues shared with me their experiences, including the warm reception extended to them by my Negril network. In 2017, I was saddened to hear of the passing of Mrs. Sylvie Grizzle, a prominent hotelier and founding member of the Negril Chamber of Commerce (NCC). In the past when I visited Mrs. Grizzle, it was more than an interview; it was also lessons in hotel management, local politics, and community planning. She took the time to tell me what was going on in this village that was now a tourist mecca.

The sense of "Jamaicaness," which implies cordial, reciprocal, and mutual respect, is at the heart of Jamaican cultural practices. I was fortunate to be the recipient of that way of being from a number of women who were interlocutors in this study. As I learned more about the inequitable encounters between Negrillians and tourists, I also recognized the practice of the "sucking of teeth" sound—to show "skin-teeth," that is, the "drawing of lips back, revealing the teeth to express a grin," but also to hide a grimace displaying signs of disdain. In the world of service and hospitality, all of these cultural practices of Jamaicaness and showing skin-teeth were noticeable social behaviors.

In December 2019, at the beginning of the "high" tourist season, Negril and the rest of Jamaica's tourist sector were poised to have a banner season. Instead, those hopes were dashed with the impact of the COVID-19 pandemic shattered those promises of success. The country closed its borders, airlines suspended flights, and the government constructed geographical boundaries to keep tourists in a northwestern corridor location while the rest of the country was locked down, closing schools and normal transactions of life. This global pandemic had significant ramifications for the local population, including their own health issues and the loss of employment across many of the sectors, particularly in tourism, the lifeline of the entire Jamaican economy. The pandemic is now part of this ethnography about the lives and livelihoods of women tourist workers in Negril.

Acknowledgments

Funding for this research in Negril came from numerous sources. Academic centers of the University of Maryland College Park answered my call. Included were the Graduate Research Board (Graduate School), International Studies, Africa and the Americas Committee, the College of Arts and Humanities, the College of Behavioral and Social Sciences, and the departments of Women's Studies and African American Studies. In 2010, the interim director, Mrs. Joan Cuffee, invited me to the Dame Nita Barrow Unit, Institute of Gender and Development Studies at the University of the West Indies, Cave Hill Barbados, and graciously welcomed and provided me with a visiting professorship. As a member of the Association of Feminist Anthropologist team led by Dr. Nandini Gunawardena, I received a travel grant from the School of American Research. Frequent flyer miles and other personal recourses made up for the rest.

Two interdisciplinary Ford Foundation–funded projects focusing on women, work, and family provided a receptive audience of women scholars from nearby universities in the Washington, D.C., area. Our collective focus was on women, work, and family issues and my research in Negril provided an international element. Later on, a grant I cowrote from the Rockefeller Foundation brought together a group of international feminist scholars of color to convene at the Bellagio Study and Conference Center in Italy. There we discussed, shared, and directed our attention on the intersections of women, work, and the global economy. My work also afforded me an invitation to participate in a UNESCO conference on tourism held in Havana, Cuba.

I was invited to give a lecture at the Latin American/Caribbean Studies Center at Colgate University, Georgetown, and Columbia University, where I received valuable comments through collegial conversation. My presentation to the Department of Anthropology, Syracuse University, held significance for me as more than three decades back I was an undergraduate major in that department. A few members of faculty who were there at my time were in the audience. Conference presentations were delivered at meetings of the Caribbean Studies Association, the American Anthropological Association, the Society for Applied Anthropology, and the Latin American Studies Association.

My colleagues in the newly named Harriet Tubman Department of Women, Gender and Sexuality Studies were immensely supportive of me and this project, particularly Debbie Rosenfelt and Seung-Kung Kim. While they were graduate students, Dr. Barbara Shaw and Dr. Katie White provided me with invaluable help and insights in organizing research materials. My colleague, Dr. Erve Chambers of the Anthropology Department at College Park, served as my guide to tourism studies. All along the way, my "sista anthros" Drs. Alaka Wali, Cheryl Mwaria, and Leith Mullings prodded and poked me to focus and to stop multitasking. Dr. Faith Mitchell, who I met in the late 1970s while we were both anthropology graduate students conducting research in Jamaica, came to my rescue in the last phase of this writing project. My Jamaican colleagues, Dr. Patricia Anderson, Ms. Marva Phillips, and the late Dr. Donna McFarlane, were always there for me. Forever I will be indebted to the brilliant scholar and eminent anthropologist Dr. Johnnetta B. Cole, whose support, kindness, and generosity of spirit I seek to emulate. The last stage of this writing project was given a boost by Mrs. Nola Stair of the Negril Chamber of Commerce.

Finally, none of these activities, or my career, would have happened without the love and support of my husband, Dr. James M. Walsh, and our now-adult sons, Dr. Shane Bolles Walsh and Robeson James Walsh. Our family now includes Stephanie Gerhold, whose expertise in the Spanish language and Latin American cultures married Shane's and now fits quite nicely into this family of international travelers and scholars.

Women and Tourist Work is dedicated to the memory of Dr. Leith Mullings, a Jamaican by birth, New Yorker by habitat, Black feminist anthropologist, influential scholar, leader, and activist.

Introduction

Women and Tourist Work is about the women who work in tourism in Negril, Jamaica. It is this collective effort of workers who greet, serve, attend to, and accommodate tourists that makes a trip to Jamaica a positively memorable one. To catch the rhythm of the island from the time a visitor sets foot on Jamaican soil, the welcoming voices of women are heard in the arrival terminal of Sangster's International Airport, the gateway to Jamaica's hot spots of tourism. The lyrical folk music and the melodic cadence of the Jamaican vernacular language of the women are heard and admired. As they direct the flow of traffic, these Jamaican women tourist workers shepherd visitors to the vehicles for transport toward their final destination on the most northwestern tip of the island. With words of assurance, the travelers are reminded by the women that Negril is just an hour and a half away, so relax and enjoy.

Emphasized here are the work lives and experiences of women tourist workers rather than tourists. This perspective flips the script of most tourist studies that are written from the perspective of the tourist or the needs of the tourist industry itself. This viewpoint of tourism is germane to understanding the centrality of tourism in Jamaica's economy and consequently its impact on the citizenry. Tourism is Jamaica's second-largest employer and the largest earner of foreign exchange for the country. More importantly, a fifth of the entire Jamaican workforce is classified as a female hotel/restaurant worker (Statinja.gov 2019). The significant role played by women in tourism is even more evident when traditional formal female employment sites such as in banking, retail, or clerical are added to this occupational mix. Women are truly the heartbeat of the tourist industry as Jamaica remains one of the top vacation destinations in the world, prominent in the Caribbean and in the village of Negril.

Women and Tourist Work captures three decades of observations, participation, and examination of tourism as it made in-roads in the very fiber of Negril with the focus on women workers who were my interlocutors. As a Black feminist ethnographer of tourism, my concern was framed by the following questions: What kinds of jobs were available to women? How did the kind of work provide a livelihood for women tourist workers in Negril? What obstacles were put in place predicated by their class (indicated by education and skill levels) and other social impediments? And what did the women think about tourism as an overall outcome of national development? During my trips to Negril in person, or through other modes of communication, I posed those questions to women tourist workers. In this way, those who earn their living in this now-burgeoning sector of the Jamaican national economy have a way to voice their opinions and concerns via this project.

ORGANIZATION OF THE BOOK

Organizing this book required a four-step process of unpacking, dissembling, reordering, and constructing. First, tourism is a multifaceted industry. Consequently, there are a number of moving parts such as the variety of service, quality of service, accommodations, face-to-face encounters, and hospitality to name but a few. Second, this industry is an ongoing process in Jamaica and ever-evolving in Negril. Third, through an analysis of the Jamaican-gendered system of labor, the critical role of women and work in tourism in Negril is best understood. Finally, together these three points serve as explanatory tools that shape women's experiences as workers in tourism in Negril, Jamaica, during this time period.

To start there are conceptional definitions and descriptions used that frame this ethnography. *Women and Tourist Work* is a study of tourism, which is a multifaceted industry where the bottom line is service. Naline Joseph, a Caribbean tourism marketing specialist, defines *service* as any activity, benefits, or satisfaction that is offered for sale. Service is essentially intangible and does not result in the ownership of anything. Its production may or may not be tied to a physical product (Kotler 1999; Joseph 2005, 172). Nonetheless, service is tourism's product, performed with smiles regardless of the difficulty, or circumstances of the task. Tourists come to Jamaica looking for enjoyment under a tropical sun where aquamarine waters wash up on beaches of white sand with coconut trees swaying in the breeze. They seek ways to satisfy their sense of pleasure and adventure, and are drawn to Jamaica by images of sand, sea, sun, and sex in their mind. These images are reproduced by targeted global marketing promotions viewed on television and in print media advertisements, partnering with travel agents in person and/or online services like TripAdvisor, and airline direct flights to the country. Whether it's a nonskilled worker who sells coconut water on the beach or the manager of a five-star hotel, the objective is to provide the best service possible to the tourist, which amounts to the measure of success in this industry.

As hospitality and human resource specialist Chandana Jaywardena (2005, ix) noted, "tourism is a 'people' business" and "good service is never an accident" but does require local knowledge of issues, understanding hospitality service, and the role of women in the Caribbean. On top of the scale of "good service" is the concept of hospitality. Hospitality provides a positive experience that will reap the benefits of a higher customer retention rate, as opposed to their counterparts who offer less-than-pleasant experience. It requires a worker to give selflessly to make that positive experience happen. The end goal is a satisfied guest who returns to repeat that experience and tells others about the benefits of that occurrence. The women whose work experiences are described in *Women and Tourist Work* recognize the importance of a satisfied guest and see this as an aspirational aim toward making a decent living and providing for their families.

As a service industry, the elementary nature of tourism is based on personal interactions. Encounters between individuals are determined by the nature of the relationship waged in terms of consciousness, identity, and intentionality. The unevenness of the playing field in tourism is sometimes predicated on the level of service rendered and the value placed on that work. For example, providing the necessary basics of housekeeping is fundamental for lodgings and accommodations. Nonetheless, the work of a housekeeper/maid is deemed of little value by the low wage that employers pay and understood as such by guests. When that work requires more than the necessary, such as the outcome of a nasty and messy guest, the labor expended by the worker can result in surliness. Those actions on behalf of the worker are far from the acquiescence usually afforded to a guest, even if she reported her disgust to her employer. For the woman tourist worker, the feeling of alienation of her right as a citizen and the powerlessness of her class status in her own country add fuel to the situation. Unless the employer recognizes the problems of this housekeeper versus messy guest situation, there will be no recompence for the former and none for the unruly guest. These kinds of circumstances continue to be an issue for the Jamaican Tourist Board, which works hard to amend these situations in their promotion efforts. Often tourist workers just feel a sense of being profoundly disrespected, which ordinary people in Jamaica feel every day (Meeks 2000, 153–54).

CULTURAL CUES

One of the cultural behaviors that help tourist workers deal with the inequitable situations they might find themselves in is to show "skin-teeth." This
behavior comes from the days of enslavement when any act outside of com-
placency was a form of resistance. Enslaved women did not take their station
in life lightly and verbal displays of displeasure were key weapons. In one of
her classic works on Jamaican folklore, Martha Beckwith (1925, 88) noted
the following, saying: "No everybody wha 'kin teet' wid you a you frien,"
and "no kin teet' a kin teet." These sayings mean not everyone who you
show a smile is a friend, and not every laugh is an honest laugh (Cassidy and
LePage 1990, 261). Showing skin-teeth, the sucking of teeth and revealing
of tight lips, is a cultural practice that hides true value of behavior from the
receptor, especially when that person is deemed not a social equal. It is easy
then to see how skin-teeth are used as a cultural coping mechanism in the
tourist industry across job categories. Skin-teeth are used against the social
inequality, racism, and sexism exhibited in any kind of face-to-face encounters of tourism.

On the other hand, women exercise on many levels what Jamaican anthropologist Don Robotham described as their "Jamaicaness." Here, there is a
shared feeling of social responsibility, a "moral feeling for each other" (cited
in Meeks 2000, 152) that can morph into a kind of public warmth felt by
tourists. Jamaicaness in this context mixes intentions as well as ambivalence
in a worker's collaboration in an industry that makes the tourist "feel alright"
according to the Jamaican Tourist Board slogan. In their line of work, personal service workers engage in other tactics besides skin-teeth. This has to
do with how service is perceived as labor that is seemingly effortless, and
something that comes naturally. Doing for others is just part of a worker's
livelihood and comes under the category of "managed heart" of emotional
work. This is when smiling is a part of the job and also disguises fatigue
and irritability. Without the attributes of the managed heart, the product—
a satisfied vacationer—would be damaged. Banishing irritation calls for
emotional labor in addition to showing skin-teeth. Arlie Hochschild (1983,
35–55) described different kinds of emotional labor in terms of acting, and
the amount of degree of acting necessary to bridge the face-to-face encounter.
Surface acting requires that the person only act as if they had a personal stake
in the outcome of some activity or event. For example, saying with appropriate body gestures "let me help you with that." Deep acting requires a person
to exhort feeling or to make use of the indirect ways of the imagination to
convey emotion. Envision such action in this statement, "Isn't that awful?
Now let's see what we can do to make this better." Then, there is institutional
emotion management whereby the institution sets the stage and the rules and

modes of appropriate behavior. The worker must use institutionally approved emotions or reactions to certain sets of activities, such as "Welcome to McDonalds, can I take your order?" with a smile.

Needless to say, in "the customer is always right" of tourism, service workers do all of the acting necessary to satisfy a guest, so they will return to Jamaica again, and again, or whatever is the current Jamaican Tourist Board slogan. Using both culturally appropriate skin-teeth, honed during the forced enslavement, and managed heart techniques of different acting methods, Jamaican women tourist workers are indeed succeeding.

METHODS OF STUDY AND LOCATION

Women and Tourist Work is based on ethnographic research conducted in Negril, Jamaica, from 1990 to 2002, intermittently through 2015, and since then via online research. A variety of online resources, historical records, newspaper accounts in *The Gleaner* and *The Jamaican Observer*, as well as descriptive analysis and the all-important participant observation were essential. The narrative accounts provided by the women themselves were key. Since 2016 there have been a few telephone calls and by 2021 calls via WhatsApp. Calls to a member of the Negril Chamber of Commerce were critical to keep abreast of what was happening in Negril. The long-term aspect of this ethnography focuses on women and their work in the temporal context in which they are found. Framed by the evolution of their community and opportunities afforded in the tourist sector, this study tries to capture women whose lives and experiences are constantly shaped and reshaped. *Women and Tourist Work* is a point on a continuum as Jamaica's and Negril's position in the tourist industry is ever-evolving and expanding. These fluctuations have political overtones. If there was a change in which political party won the national election, then there were shifts in plans and directives coming out of the new leadership at the Ministry of Tourism. These shifts in the direction of the industry had consequences for the women who work in tourism.

WOMEN AND WORK

In the chapters of *Women and Tourist Work*, I contend that tourism in Jamaica is a predominately "female" industry in terms of the product and the workforce. Tourism is the major source of these women's livelihoods and, subsequently, their households including their children depend upon the sector. Consequently, the precariousness of this business hindered by hurricanes, global economic downturns, as well as shifts in political largess hovers over the monetary opportunities of the women in Negril. Often, women's overdue material gains were overshadowed by gender systems that continue to be inequitable in terms of the status of women

GENDER SYSTEMS

Despite the cultural importance of women as mothers and providers, institutionally there is the basic belief of female subordination that holds up gender disparities that still grip the Caribbean at large and are reflected in tourism. Caribbean feminist scholar Eudine Barriteau (2002, 222) used *gender* referring "to a system of social relations through which women and men are constituted and through which they gain differential access and are unequally allocated status, power and material resources." As a service sector, tourism is deemed low in prestige because it is assumed that the workforce—mainly women—already knows how to perform the tasks required by jobs in the industry. Furthermore, the majority of jobs in tourism are labor-intensive. This means the sector requires a high ratio of employees to paying customers; people who come as tourists need and expect a lot of services. The kinds of jobs typed as labor-intensive are also unskilled, low-skilled, and low cost in terms of wages and benefits. The jobs in the tourist sector are viewed as the ones not only that women know how to do but also that come "naturally" to them (Timothy 2001, 246). Therefore, jobs such as housekeeping, doing laundry, cooking, serving, and so forth are female-dominated and essential but also with low economic value. Consequently, the very nature of tourism as a service sector exhibits characteristics associated with a specifically gendered segmented labor market.

On the other hand, while the majority of women employed in this sector are found in the no- or less-skilled level, there are women tourist workers who are employed in a variety of occupations that require a secondary education and/or higher education, technical skills, and advanced training. Some of these jobs are in accounting, bookkeeping, hotel management, recreation, and medical personnel. While those are identifiable female labor categories, there are still a number of women in Negril who hold jobs usually held by men. Counted among the nontraditional jobs held by women are bank managers, food and beverage managers, chefs, and head accountants. Unfortunately, the questions concerning equitable pay issues cannot be addressed in this chapter.

Nevertheless, answering the question, "What does it take for a woman to succeed?" required a methodological framework that helped to decipher the relationship of social barriers such as color and class in this Jamaican context. Intersectionality as a method/viewpoint conveys the sense that individual identity and social life are produced by multiple, overlapping, and contradictory systems of power that operate simultaneously (Kim and McCann 2010, 1422). Intersectional thinking was first developed as a way to grasp how race (U.S. Black and Brown) and gender are overlapping vectors on the social inequity scale in the U.S. context by legal theoretician Kimberlee Crenshaw (1989). For an intersectional approach to be "travel ready," it must be considered as a scaffold that exposes the nexus of social differences in specific locations. This scaffold must be constructed of an appropriate historical framework, an understanding of the differences of the axis of power, and an introspective understanding of the power of persuasion as it demonstrated in that instance (Bolles 2015, 69). Therefore, gender "a system of cultural identities and social relationships between females and males as a variable in any study of human relationships" (Swain 1995, 247) must be complicated by how a series of other factors influence those relations, thereby underscoring the true value of an intersectional perspective. In this way, the long-term history of economic gender disparities and the importance of women's earning and providing in the Jamaican context coalesce with the central factor in the development of what is determined as a gender ethos. An intersectional approach reveals how this gender ethos is enacted across those lines of class, color, and other points of difference.

This philosophy is embedded in the socialization of girls, who learn earlier on at home, and sometimes at school, that it is important to be economically self-reliable as much as possible and to provide for their families. At one point in time, particularly for the middle class, reliability came through heterosexual marriage. Now, there are more opportunities available for middle-class women than before so they are not reliant on marriage for economic viability. However, for those dependent on working-class women's wages, the intersection of social factors such as class, color, training, and education yields a percussive effect and remains the major topic for this study on tourism in Negril, and in general that of social science research. Thus, when opportunity is at the front door, with appropriate skills and self-confidence, women must seize the moment. This dictum is very pronounced in the growth of the kinds of roles that women play in Negril's tourism, as this work recounts.

Women and Tourist Work is the written account of descriptions of the scenes and encounters that the ethnographer witnessed and interviews conducted, coupled with historical analysis of the arena of research, and then framed by theory or a series of theories that become explanatory tools for readers, usually other scholars and students. Further, there is special value in the work when it is shared with those who are represented on the page in a reciprocal and respectful way. This is a hallmark of Black feminist anthropology and those who follow decolonizing methodologies that consider how power operates differentially (see Davis and Craven 2016, 13).

A benefit of ethnography as a method of inquiry allows a viewpoint that historizes a place and a group of people as it positions them in a moment in time. Moreover, feminist ethnography features essential elements that not only pay attention to space and time but also focus on the values of reciprocity and respect that should emerge in the face of the differences between researcher and the interlocutor (Davis and Craven 2016, 11). Such is found in *Women and Tourist Work*, which provided a frame that shares what ethnography came before it and then moves onto the accounts of the study. In the 1990s, when this research in Negril began, time was spent living and learning about how women workers made a living and made sense of the world around them. Such are the methods of participation and observation that yield awareness and appreciation of what was occurring in the community. Living, traveling, eating, shopping, working, and generally participating in the day-to-day life of the people we study is the hallmark of ethnography (Leite et al. 2019, 2). All the while, the researcher was examining her own experiences and understanding of the social, political, economic, and other bodies of knowledge that helped to explain what unfolded in "the field" setting.

The goal was to look at the range of workplaces and economic activities occupied by women, noting differences that surfaced in everyday occurrences during the fieldwork in Negril. Since the goal was to talk to a wide range of women workers of diverse occupation, age, education, class, and racial and ethnic groups in their work settings, it required visits to multiple sites where interviews were conducted with a wide variety of women tourist workers. In her research on tourism in Cuba, Kaifa Roland (2011, 112) remarked that by participating in daily activities, "researchers not only gain trust while being instructed to the correct way to do things, they, also have the opportunity to better communicate the meanings of those cultural practices to others in the resulting ethnography." Subsequently, there were a lot of informal conversations on the side of the road, visits to small- and medium-sized hotels and lodgings, and after-hour sojourns with those whose workday started just around midnight. Formal and informal interviews were guided by a survey of topics engineered to follow the research agenda, and copious notes were written. In sum, the methodology applied in *Women and Tourist Work* provided both descriptive and analytic approaches that illustrate how the differences in the lives and experiences among women tourist workers came about. All of this was made clear by an intersectional perspective that enabled notions of reciprocity and respect between the researcher and the interlocutors, thereby undergirding a Black decolonized feminist ethnographic praxis. The next step reviews some of the pertinent scholarship on tourism, particularly coming from the Caribbean as a source of guidance for this project.

CONTRIBUTIONS TO THE STUDY OF TOURISM

It was only in the 1970s that leisure activities, including tourism, began to receive the attention of social scientists. As pioneer of the anthropology of tourism Valene Smith noted, in the decade between the first edition of *Hosts and Guests* (1978) and the second edition (1989, x), "No one now doubts its [tourism's] importance or questions its future as one of the leading world industries." Beforehand, tourism was most productively viewed not as an entity in its own right, but as a social field where many actors engaged in a complex interaction across time and space, physical and virtual (Leite and Graburn 2009, 37). The increasing awareness of this social phenomenon came from the rise of mass tourism among the world's increasing numbers of middle-class populations. Sharon Gmelch (2004, 7) suggested that the academic community did not take the study of leisure, sport, and tourism seriously as topics worthy of scholarly pursuit. In terms of anthropology, a discipline that historically focused on peoples of the global south (developing world, former colonial world), Bruner (2005), G. Gmelch (2012), and Chambers (2010) all agree that there was reluctance on the part of members of the discipline to see that the people with whom they studied in fieldwork settings were becoming the objectified "exotics" of the tourist trade. When anthropologists turned to study tourism, it came in the form of ethnographic accounts of specific tourist places and situations (Chambers 2010, 2).

Since 1492, the Caribbean region garnered a sense of history of heightened worldwide orientation. The rise of global capital, fueled by the outcome of forced African enslaved labor, indentured subcontinent Indians and Chinese, and others, afforded massive wealth for most wealthy Whites and Creole peoples and to colonial governments (see Shepherd 1999). The confluence of money, power, race, and class differences was so very much pronounced in the field of the region's literature exemplified in Guadeloupean novelist Maryse Conde's *Tree of Life, a Novel of the Caribbean* (1992) and historian poet Edward Kamau Braithwaite's seminal writing *The Arrivants: A New World Trilogy* (1973). In *The Tree of Life*, Conde's generations of characters move across the Caribbean, to San Francisco, and to Paris in search of work opportunities as immigration allows but clouded by colonialism and racism. Barbadian-born Braithwaite's trilogy of poems, *The Arrivants*, highlights the richness of beauty and wealth, as well as the violent past of the Caribbean, European expansionism, and migration to the metropole and then to Africa and back. Clearly, literary writers are interlocutors of social and economic transformations witnessed as Caribbean people and conveyed in their art as citizens of the region. Artistic expressions formed a political critique and social commentary, as any devotee of Bob Marley lyrics would acknowledge, and can be found in Bajan Jack Deer's biting calypso "Jack" (see Gmelch

2012, 23). The past and the present exemplified in literature and music are predisposed in the ethnography of the region. Nation-building, now not a postcolonial enterprise but a neoliberal convergence of private investment and public sector facilitation, serves as an explanatory lynchpin of local Caribbean people whose homes are now tourist sites.

The following discussion features four anthropological ethnographies carried out in the Caribbean. Each center on global tourism as a connection to understand culture, identity, and economics. The studies focus on British Virgin Islands (BVI), Carricou (Grenada), Boco del Toro (Panama), and Barbados.

Based on twenty years of traveling to BVI, Cohen (2010) provided an "outsider/insider" perspective where she looked at a society that is a British Overseas Territory, its citizenry British subjects, and the currency is the U.S. dollar. The importance of the work rests on the impact that tourism had on society and national identity. Anthropologist Collen Cohen combined a close reading of her observations and encounters with BVIslanders through popular songs, festivals, and migration flows that illustrated how this society, an "archipelago of islands," turned its eye toward a neoliberal model and slowly transformed itself into an ideal tourist spot with a focus on yacht tourism and off-shore finance. Of concern was how this form of social and economic development based on global forces influenced national identity formation. The notion of whose paradise was in contention—that of the tourist or the dispossessed islander—was the major theme of the work. Cohen (2010, 4) wrote of her concern with

> the effects of tourism and tourist desire for paradise upon the people who live in the BVI. A good many of these people are occupied with servicing tourists and satisfying tourist expectations, but even the lives of BVI residents who do not engage directly with the visitors are shaped in multiple and complex ways by tourism.

Identities of "belongers" based on claims of BVI citizenship and "nonbelongers," who include laborers who migrated from Jamaica and elsewhere to work in the BVI tourist sector, were thoroughly examined. Further, tourists' expectation/fantasy that the BVI remains one of "nature's little secrets" undermined the realities of the physical and cultural transformation of these islands into tourist developments. One of the ways of maintaining "belonging" were artistic expressions found in music, art, dance, and festivals as demonstrative of BVI national identity all the while serving as a source of entertainment for both "belongers," "nonbelongers," and tourists. The notion of who belongs is echoed in another text on the Caribbean and tourism.

The construction of a tourist paradise aided and abetted by neoliberal schemes fostered by private concerns and facilitated by governments was examined by Carla Guerrón Montero in her work in Grenada (2010) in conjunction with her long-term projects in Panama (2020). Her comparative study of two islands focused on Carricou—the companion island of Grenada and Colon—and Bocas Del Toro—islands off of the mainland of Panama—each providing a way to see how "commodity plantations become tourism transplantations." What appeared in this examination was the connection of private investment with government backing in the development of tourism on both islands, but more so the difference in the response of the African-descendant populations whose homes and cultures housed tourists in hotels that they built and who are in service to those visitors. Kayaks, the name Carriacouans call themselves, hold a major cultural event called the Big Drum Dance. The event became a tourist attraction, but its authenticity was gauged by where it took place—at home in the village or in front of a paying audience. In this way, the Kayak community continued to value their history and culture all the while maintaining a way that its culture "will not 'suffer' or be diminished by tourist development" (2015, 1). In contrast, following a tolerance promoted by the government, Panamanian African Caribbeans, descendants of the nineteenth-century Panama Canal diggers, claimed a "Jamaicaness" was actually a regional Caribbeaness based on their phenotype (skin color, hair texture, posture) and cultural displays of "Jamaican" performativity that are acts of inventing and "reinventing traditions for tourism consumption" (2010, 29).

Guerrón Montero continued her examination of nation-building tourism in *From Temporary Migrants to Permanent Attractions: Tourism, Cultural Heritage, and Afro-Antillean Identities in Panama* (2020). Based on long-term ethnographic and archival research, she again addressed how public and private transformative forces fueled by tourism constructed a national identity that embraced multiculturalism as an attribute to the national brand. No longer a country of indigenous, Creole, and colonial Spanish descendants Panama is now embracing Afro-Antillean as part of their multiculturalism and nation-building. Guerrón Montero analyzed the ways in which tourism became a vehicle for the development of specific kinds of institutional multiculturalism and nation-building projects in a country that now prides itself on being multiethnic and racially democratic.

Guerrón Montero focused on Panamanian Afro-Antilleans who reside in Bocas del Toro. Previously this group struggled to have their presence acknowledged in Panama's social and cultural history, first as diggers in the construction of the Panama Canal and then as socially disparaged plantation workers for the United Fruit Company with their tropes of place assigned to side-bars in Panama's narrative of nationhood. Through careful archival research, Guerrón Montero marked how Afro-Antilleans' presence morphed from Black-skinned plantation workers to Caribbean tourist workers. She noted that public opinion/government recognition was facilitated through tourism. Furthermore, Afro-Antilleans seized the moment to improve their status within Panamanian society simultaneously identifying with their Caribbean heritage in ways that still counter their existence in the framework of national identity.

Like many Caribbean countries, the development of tourism in Barbados began in the 1950s–1960s and was full-blown in the next decade. As George Gmelch (2012, 7) remarked, "Spurred on with government incentives, hotel construction boomed, unemployment dropped, and visitors began to arrive." Rural Barbadians who previously worked in agriculture were now employees in some form or fashion in the various subsectors of the tourist industry. As of 2019, 40 percent of the Barbados economy was generated by tourism, and two in five jobs were created in the sector, according to industry information. Anthropologist George Gmelch's *Behind the Smile* (2012, 2nd edition) structured a view of tourism that began with a social historical framework of Caribbean tourism that pivoted down directly to Barbados as an example of now private investment and government incentives work hand in hand. Further and more importantly is the ideal of the smile, the genuine and ingenuous way that Barbadian tourist workers engage in their encounters with visitors to their island. Twenty-one narrators, five of whom are women, shared their stories of their daily and work lives in the sector. Arranged by the location, such as the airport, the hotel, the beach, and the attractions, the fifth section focused on how the government, the Barbados Hotel, Tourism Association, and the like assess outcomes of the industry. Each story conveys how each worker viewed their encounters with visitors, and what tourism meant monetarily to the country and to them as Bajan citizens. A teller and money exchanger at the Grantley Adams airport remarked, "I believe you should always be as helpful as you possibly can. . . . It's nice to be important but more important to be nice" (Gmelch 2012, 50).

Featured first in the 2003 edition of *Behind the Smiles*, the narratives of the twenty, now twenty-one, Barbadian tourist workers were based on interviews conducted with a team of George Gmelch's undergraduate students from Union College, New York. What a great team it was. Under the guidance of their anthropology professor, the outcome of that summer work was an extraordinary blending of expertise and the newly found wonder in the power of everyday wisdom coming from ordinary people. The epilogue speaks directly as to how these young people were affected by this experience in terms of their own positionality of class, race, and gender. It also allowed them to think more deeply about how tourist development impacted their own daily lives back in their own backyards. In Barbados, by centering the work on the voices of the people engaged in tourism, the interviewers recorded the workers' own sense of agency as evoked in their actions on the job and in their households. As a bartender commented, "When people come to the bar, they do not want to hear your problems. They come to tell you theirs. . . . So, when I talk to them, I'm always happy, laughing, ha, ha, I never let them know that I've got problems too" (Gmelch 2012, 70). The centrality of the workers' experiences in tourism plus the way the ethnography was structured made this book a "must read" in anthropology and in business and management studies.

These four studies of the Caribbean used the prominence of tourism as the vehicle for change for those who were their interlocutors as well as the islands in which they were located. Although identity and sense of belonging were major factors for these works, the impact of a tourist product and the significance of tourism in the lives and livelihoods were most pronounced. In a similar fashion, what occurred in Negril has resonance to those Caribbean tourist studies.

FINDING OUR VOICES

In the beginning, part of the research process included reading countless studies on tourism, and exploring what planners, economists, the tourist industry, novels, films, and anthropologists and sociologists had to say about the subject. Jamaica's tourist industry had a long history but was quite simplistic in the way women workers were addressed. Also stymied in these early research efforts was the fact that the Jamaican Government (GOJ) had not disaggregated its data on tourism by gender, occupational type, or cross-referenced it with other sectors, such as construction, communication, and transportation, which all contributed to the development of the industry. A reoccurring question asked, "Where were these 'invisible hands' that make things possible for the tourist?" To answer this question, this research circled back to the comfortable turf of fiction.

A Small Place (1988), a small jewel of a novella/essay by Antiguan-born novelist Jamaica Kincaid, came to mind. This piece of decidedly nonfiction prose calls attention to the social and economic ironies of political independence faced by a small island state in the Caribbean. Kincaid wrote how the very places that tended to attract tourists were often the sources of difficulty for those who lived there. For instance, for the tourist, clear, sunny skies meant more time for fun in the sun, while for the local population those skies meant lack of rainfall, making fresh water a scarcity and a precious commodity. For tourists, however, the beauty was all that mattered—the drought was someone else's problem. There are contrasting perspectives between what the tourist sought and was provided versus what local people were denied. In the end, the exotic and often absurd misunderstanding that tourists have about this "strange" culture, and its political and social environment, kept the tourist from really knowing the place they went to visit. *A Small Place* had a great impact on Antigua, so much so that the author assumed a nom de plume, "Jamaica Kincaid." The book made her family so angry that she had to change her name, thereby disassociating herself from their criticisms. For a while, the government of Antigua and Barbuda made Jamaica Kincaid a persona non gratis.

In addition to the brief acknowledgment of influence of novels, poems, and music in the writing of *Women and Tourist Work*, the power of the Black poet/essayist June Jordan's "Report from the Bahamas" (1985) proved to be most profound. This essay was written as a personal narrative as she examined the people she came across while on holiday in the Bahamas. Jordan encountered Black Bahamian women such as the hotel maid named "Olive" to the old women street sellers hawking trinkets with whom she believed she had affinity. She wrote,

> I notice the fixed relations between these other Black women and myself. They sell and I buy or I don't. They risk not eating. I risk going broke on my first vacation afternoon. We are not particularly women anymore; we are parties to a transaction designed to set us against each other. (2010, 161)

Jordan's essay exposed the possibilities and difficulties of drawing alliances based on perceived oppression, race, and gender identities, but as she surmised, the differences made the connections too hard to overcome. Jordan as well as the studies just mentioned centered on power dynamics and discourses that affirm the existing imbalances in human international relations and thus constrain and limit the life chances and opportunities of millions of women and men (Pritchard et al. 2007). They go on to say that the tourism industry has the potential to offer opportunities, but yet can be a force for "ghettoization, oppression and inequality thereby making indigenous women and female employees as part of the tourism package" (9). This is a good warning.

MORE ON ANTHROPOLOGICAL TOURIST STUDIES

Feminist anthropologist Florence Babb (2011, 9–12) wrote that she found inspiration beyond anthropological tourism studies to those of cultural analyses produced by postcolonial scholars. Particularly important to Babb was Edward Said's (2003) influential explorations on "orientalism," which examined the power infused in the machinations of colonialism, and the work of Mary Louise Pratt (1992), which identified encounters and processes whereby representations from the global metropolis were deployed by certain groups in the periphery. Nonetheless, there is no doubt that anthropological studies have made significant contributions to the study of tourism developing from a basically negative judgmental stance into a sophisticated and nuanced understanding of how tourism influences culture and society (Graburn and Moore 1994). The chapter authors and editors in *The Ethnography of Tourism* (2019) not only examined the growth and depth of this line of study and its history but also lauded one of the major figures in the field, Edward Bruner. Cited in the introduction (Leite et. al. 2019, 11) was Bruner's articulation of a revised concept of culture that referred to it as "'experience,' to emphasize what is meaningful in people's lives. By 'experience' we meant what emerges to consciousness, as opposed to what people say or do." Chambers (2010, 3) argued that

> culture is expressed by the ways in which members of a group determine and symbolize the meaningfulness of their lives . . . Tourism, with its multiple realms of human interaction, provides ample opportunity for the play of cultural processes and for the invention of new forms of cultural expression.

Chambers' review of some of the trends found in the literature allowed a categorization of this body of knowledge. This perspective was fundamental in *Women and Tourist Work* as a way of seeing tourism as a mediated activity displaying a variety of interventions and an equally diverse array of interpretations by the tourist, the tourist worker, and the anthropologist. Babb (2010, 21) remarked that we (anthropologists) should not just write off tourism as a global industry that rides rough shod over cultures but rather should examine the complex ways in which it is negotiated by all involved. Further, the influential effect tourism plays in communities can truly expose the "stark differences" in economic and cultural opportunities that result of this expansion. Mimi Schiller's *Consuming the Caribbean* (2003) explored the myriad ways that Western European and North American people have "unceasingly consumed the natural environment, commodities, human bodies, and cultures of the Caribbean over the past 500 years." The agency of the peoples of the region is not the main ingredient in Schiller's volume. The attention still rested on "the Caribbean" as viewed from those outside who consume and commoditize Caribbean culture, objects, and subjects (Schiller 2003, 150–55). Here the "tourist gaze" is defined as the tourist encounter that decontextualizes a site and imposes a narrative claim upon it that objectifies the sociocultural space and people who live and work there (DiGiovine 2010, 8). Consuming and commoditizing the Caribbean and other destinations reveals that tourism is a force to be reckoned with in terms of its impact on a society and its inhabitants. This sample of anthropological critique of tourism demonstrated the value attached to this kind of work, which must be done with due diligence, care, and in a way that is mindful of those whose lives are represented.

CONTENT AND STRUCTURE OF THE BOOK

Women and Tourist Work begins with a discussion of the history of Caribbean tourism with major emphasis on Jamaica. After the introduction of airline flights directly to Montego Bay, Jamaica took the lead among the English-speaking countries that were using tourist-sector investments as a course in economic development and growth. These occurrences were also prompted by the 1959 Cuban Revolution when U.S.–owned hotels and resorts were nationalized and the economy took on a socialist format. The result was a redirection of investment from Cuba toward Jamaica. Innovation in the tourist industry came with the development of the all-inclusive model championed by the Jamaican Issa family and Jamaican appliance trader-turned-hotel-mogul the late Gordon "Butch" Stewart. In addition, all-inclusive resorts came with other international investors coming from Europe and the United States. Private investors' name-branding and efforts earmarked by government policy promoted not access to a tourist experience, but the quality of the visit. The key to success was the importance of service, hospitality, and a sense of professionalism that ensures the value of the tourist experience and repeat visitors to Jamaica.

The second chapter of the book focuses on the development of tourism in Negril. The history of Negril would not be complete without a discussion of the essential Negril landscape, the beach. Negril had a late start as a tourist spot, as former fishing village and coconut producer, seasonally tied with the nearby Frome sugar plantation area. The tall coconut tree blight that took down the trees resulted in wide stretches of pristine white sand beaches that washed against crystal clear aquamarine water, becoming a main attraction for tourism. With a late start in the 1970s, there continued to be wide range of accommodations available in Negril that made this vacation locale unique. With this special mixed blend of small cottages that are family-owned, small hotels, medium-sized accommodations, and all-inclusive resorts, these spaces are located on stretches of beach, cliffs, and on each side of Norman Manley Boulevard (the beach road). Each of these properties has histories of their own that contribute to the distinctiveness of Negril, especially with kin and family ties of ownership that surface in community relations. Negril has concerted efforts in preservation of the morass swamp, the coral reef, recycling projects, and community-based work. Particular to 2020/2021 are the determined public health plans to contain the spread of COVID-19.

The center of attention of chapter 3 is women, work, and family. Using an intersectional gendered system, analyses begin with a history of women first as members of the enslaved community. It then charts how British colonialism dictated a gendered system based on race/color and class that emerged out of slavery into schema of social inequality. Of significance, in Negril's tourist sector, women were not confined to specific female employment arenas. True, the majority of women work as housekeepers, maids, and food servers. However, in Negril women are also bookkeepers, accountants, sous chefs, activities coordinators, and operators multilingual tour desks. The discussion then frames how some women were able to circumnavigate social obstacles. Included too is how women used modes of cultural practices that emerged from the days of slavery. First were efforts of resistance and to show "skin-teeth" as a mechanism to cope with racism, classism, and misogyny on the part of those whom they served. The second was the use of a sense of Jamaicaness whereby mutual respect is given when due and valued as a sign of endearment. All of these actions were framed by managed hearts and service without servitude.

The next three chapters are based on location and therefore the women's work sites. Chapter 4 focuses on the beach, where ethnographic practices of participation and observation earned interviews and encounters with women interlocutors whose workdays are found on the sandy beach. The beach was the work site of women workers as vendors and hair braiders who provided formal and informal face-to-face encounters with one another and with tourists. This codependency fueled the precarity of the women, who trod the beach hawking their wares, often evading hotel security, who discourages this activity for the sake of tourist comfort.

Another distinctive feature of Negril is the bevy of women entrepreneurs who applied their business acumen in the tourist sector. Chapter 5 directs our attention to a number of women business owners of varying sizes. The model of achievement is the history of the higgler, a microbusiness woman emerging from the days of slavey; the modern version now faces competition from international hotel chains. There are extra hurdles that small businesses owned by women must overcome. Self-employed women in various kinds of activities are also discussed.

The final ethnographic is chapter 6, where the workday setting occurred at nighttime. The nightlife centers around Negril's offerings of food and entertainment. Dining, dancing under the limbo pole in a resort, a fashion show, and "OTJ" (on-the-job training) are all a part of the lure of a vacation in Negril and the women worker who makes this all happen. Extraordinary service is the premium and the aspirational goal of these women.

The concluding chapter brings together the pieces of tourism: of service, neoliberal investment in the tourist industry that engaged the public sector as a purveyor of cultural goods. Negril has grown from a small spot on the Jamaican map, to a major tourist spot that made a special niche because of its late arrival to the tourist business. However, of great importance are the voices, experiences, and work lives of the women who participated in this study. This conclusion attempts to address the following: How has the gendered structure of tourism afforded women a decent wage, respect, and sense of self in this global marketplace of tourism? What was the exacting toll of emotional labor, mediations, and other experiences on women in this industry? Given the toll of COVID-19, those answers are all posed as conditional responses. *Women and Tourist Work* is the result of more than twenty years of research, personal engagement, and online and face-to-face encounters all tempered by Black feminist ethnographical praxis. Jamaica continues to be the Beat of the World.

Chapter 1

Brief History of Caribbean and Jamaican Tourism

The history of Caribbean and Jamaican tourism connects to European and British colonialism, U.S. hemispheric power, decolonizing government efforts following World War II, and postcolonial economic developments. *Women and Tourist Work* begins with a discussion that positions tourism in a wider context of the world, the region, and then narrowing down to Jamaica.

TAINO HOSPITALITY

The Caribbean was the first land sighting by Christopher Columbus in 1492 when he happened upon what is now the Bahamas. A few months later, on or about December 24, 1492, Columbus was temporarily shipwrecked off of the coast of Hispaniola (modern-day Haiti/Dominican Republic). Historian Franklin Knight (1978, 6) cited the first encounters between the Spanish visitors with the indigenous Arawak, now referred to as Taino. Columbus chronicled:

> The King (Arawak) sent all the inhabitants of the village out to the ship in many large canoes and in short time we had cleared the decks. . . . Nowhere in Castille would one receive such kindness or anything like that. . . . Such was the help that this king gave us. . . . There are no better people and no better land in all the world. . . . They love their neighbors as themselves and their way of speaking is the sweetest in the world, always gentle and smiling.

By 1509, less than twenty years later, the kindness of the Arawak was rewarded by deception and their almost annihilation as a people through genocide, exhaustion, and disease at the hands of the Spanish to whom they

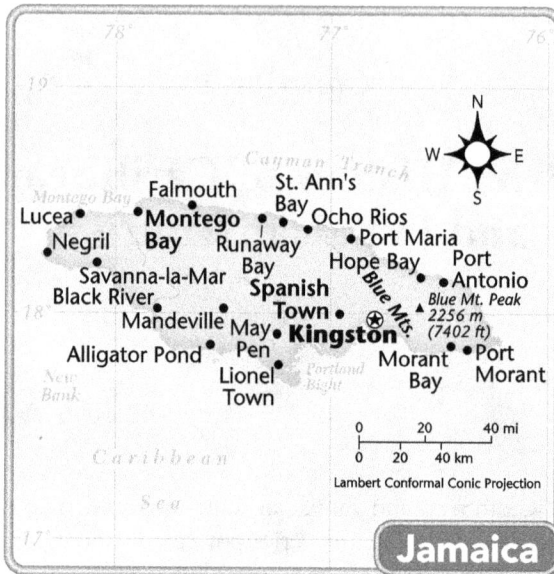

Figure 1.1   Map of Jamaica in the Caribbean. © PeterHermesFurian/iStock/Getty Images Plus/Getty Images.

initially extended their hospitality. This is not a story of the beginnings of the leisurely travel associated with Caribbean tourism. However, it acknowledges the importance of hospitality in face-to-face encounters between Caribbean people and visitors that are critical to modern-day tourism.

EUROPEAN EMPIRES AND TOURISM

The history of travel outside of an individual's locale for leisure was noted as early as 1500 BC, while later on Romans took spa treatments in the coastal waters of their empire (Casson 1994, 3). During the Middle Ages, travel to religious shrines by pilgrims across Europe and Asia was well-known. Post-renaissance upper-class European young men journeyed on grand tours to view the antiquities of Greece and Rome to complete their education. Advances in transportation, such as when rail transit developed in 1840, allowed nobility and landed elites as well as the burgeoning middle class to take leisurely trips. These trips became important social and cultural by-products and facets of life for those who could afford them during the indus-trial revolution. Anthropologist Erve Chambers (2000, 12) suggested that before the British entrepreneur Thomas Cook opened up his travel agency in 1841, there were others before him. What Cook did was to add links to the innovations of modernity—transportation and media—of the day. Further, Cook's success helped to nudge the hospitality industry into the ideology of capitalism by demonstrating that its business could be efficiently organized and managed given sufficient capital investment and operated on a scale never before imagined. New markets for tourism, catering to the whims of travelers, and reinventing places where tourists might be directed were novel approaches instigated by Cook and his followers (Chambers 2000, 13). In 1905, Guyer Feuler defined *tourism* as a phenomenon unique to modern time that was dependent on the people's increasing need for change and relaxation. Echoing Thomas Cook, the definition addressed "the growth of commerce, and industry, communication and transportation tools that are becoming excellent" (United Nations World Tourism Organization, 2021). Wealthy Europeans flocked to their Caribbean empire locations that offered warm weather year-round, beautiful landscapes, and novel experiences. The French vacationed in their colonies of Martinique or Guadeloupe, the Dutch went to their colonial holdings of Curacao, and the English ventured to their territories in the British West Indies. Those from the United States tended to follow the British or, after the Spanish American war, made their way to Puerto Rico and Cuba. Since these travelers were well-off, they stayed for substantial periods of time from short stays of two weeks to months (Gmelch 2012, 3). For example, many of these well-to-do Brits vacationed in Barbados for the medicinal powers of the "ozone" that vaporized off of the sea and engaged in other health-giving properties such as sea bathing, but not sitting on the beach for the sun. For these affluent U.S. and European visitors, the Caribbean was safe from disease and pests, they spoke their language (English, French, Dutch, or Spanish), and it was exotic enough to be a foreign place (Gmelch 2012, 5).

At the end of the nineteenth century moving forward in time, modernity, meaning "advances in technology," advances in media, and stable colonial governance, was central to the maintenance of colonialization of the global south including the Caribbean. In all facets, colonialization solidified two views of modernity. On one side were the European colonizers who were powerful rulers who socially, politically, and economically oppressed colonial subjects and exploited the resources of their colonies. Although the secession of the slave trade (1807) by the British curtailed the business of human trafficking, Africans and their descendants were still the workforce primarily on sugar plantations. Emancipation of the enslaved freed all in the region except for those in Cuba, who waited until 1886 for their freedom. Further, regardless of their independent status, Haiti (1801) and the Dominican Republic (1844) continued to be exploited by France and Spain. Moreover, the United States controlled the finances of both Haiti and the Dominican Republic under the veil of the 1823 Monroe Doctrine. On the local Caribbean side, the beneficiaries of the economies of the region were local elites and local colonial-governing politicians whose class and skin color privilege maintained the pervasiveness of the dominant Eurocentric sociocultural ideology in all social and cultural institutions.

The media of the day promoted the Caribbean region for well-to-do travelers as well as for local elites. Literary critic Leah Rosenberg (2010, 46) suggested that at the turn of the twentieth-century. Jamaican literature "had been hijacked by tourism." To prove her point, she examined how the role of "tourist writing," meaning advocating wonders of tourism and the Jamaican tourist market, was exemplified by the early work of Herbert de Lisser, who is acknowledged as one of the founders of Jamaican literature. De Lisser was writing tourist narratives by 1899, years before the publication of his first novel, *Jane Career*, which appeared in 1913. De Lisser was the most prolific Jamaican writer at the time, proving that publishing literature was a way to make a living. A middle-class "Brown" man of Afro-Jewish descent, his career was in journalism and by 1904 he was the editor of *The Gleaner*, a prominent tabloid that represented the interests of the near-White local elite. Over four decades as editor, de Lisser incorporated "tourism into his aesthetics and tourists into his business model" (Rosenberg 2010, 46). De Lisser used his tourist writing as a way to confirm the status of his local middle-class readership, implying that they were modern, cosmopolitan "tourists" in their own country by counter-positioning the quaint exoticism of the peasantry in their daily lives. Rosenberg suggested that De Lisser represented that market women, who served simultaneously as icons of Jamaica's distinctive culture, industry, and independence, were also symbols of the island's exotic primitivism. This representation of peasant women (the majority at that time) in Jamaica was embedded in tourist writings. By 1919, De Lisser was firmly established at *The Gleaner* and maintained the empire/colonial ideals of modernity. The primitive/modern binary was embraced by his local readership and enfolded in the imagery of tourists.

Krista Thompson's *An Eye for the Tropics; Tourism, Photography, and Framing the Caribbean* (2006) used the term *tropicalization* to delineate how certain ideals and expectations of the tropics informed the images of some Anglophone Caribbean islands. Thompson's concept of tropicalization described the complex visual systems through which the islands, particularly those of the Caribbean, were imaged by photographers for tourist consumption. Veiled were the social and political implications of those representations on actual physical spaces of the islands and their inhabitants (2006, 5). Nonetheless, the term *the tropics* characterized the environment of a geographic location (Tropics of Cancer and Capricorn) that received the greatest intensity of direct sunlight on the planet (2006, 6), thereby making it both exotic and otherworldly from a North American or European perspective. Thompson's focus was on the photography and postcards that were originally oriented toward tourism of Jamaica and to the Bahamas. She noted that there was "the desire and will of 19th and 20th century photographers to proffer the Caribbean as a resolutely tropical and exotic spaces" (5). Furthermore, photography, a technologically advanced tool, captured images intended for a viewer. Thompson provided a quote from Jamaica's *Daily Gleaner*, January 18, 1901, to make her point. It is paraphrased here: "They came ashore in the spirit of explorers and seemed quite disappointed to find we wore clothes and did not live in the jungle. . . . Oh, who would be a tourist with the tourists' stance, a guidebook in his pocket and a Kodak in his hand."

CARIBBEAN TOURISM TO WORLD TOURISM

In the 1920s, travel to outposts of the empire was secured by advances in technology, transportation, and communication, further increasing the ability of the affluent to tour. Joining the affluent were members of the upwardly mobile middle class who were able to engage in this leisurely affair. In the Unites States, 1920s credit was used for 90 percent of purchases, allowing for splurge spending (Altexsoft 2019). Standardized work schedules made most outbound destinations to the Bahamas (U.S. prohibition of alcohol was an incentive) and Jamaica a treat. For the middle class, who did not have the luxury of personal transits or knowledge of a second language as did the well-to-do-traveler, travel guidebooks and pre-paid group travel made their trip successful adventures (ibid.). Heretofore, the sun was to be avoided, but now sunbathing became a symbol of leisure, a luxury that the only the rich could afford (Patullo 1996, 9). Between the two wars, visitors rented or bought houses similar to the old plantocracy. Movie stars hosted parties on their own island or home as did Ernest Hemingway outside of Havana, Cuba. By the 1950s, tourism was Cuba's second largest earner of foreign exchange, behind sugar. The island welcomed more than 300,000 tourists a year, most of them American (ibid.).

Tourism in the Caribbean literally took flight as direct airline connections to Havana, the Bahamas, Jamaica, and elsewhere in region became easy to arrange and organize. Access to "the islands" for the burgeoning middle class was no longer a providence of the affluent. As Polly Patullo (1996, 9) described it, "The scramble for tourist dollars broadened and deepened." However, in 1959, the Cuban revolution caused a shift in the Caribbean tourist market. Cold War politics forced American tourists who otherwise would travel to Cuba to chose Puerto Rico, the Bahamas, Jamaica, or Barbados as their destination.

Mass-market tourism brought hotel chains to the region, enticing both American and Canadian vacationers. By the end of the 1960s, with Jamaica, Trinidad and Tobago, and Barbados gaining their independence from Britain, the growth of Caribbean tourism had unbridled influence on those economies. Caribbean intellectuals questioned if government support of tourism aided and abetted by foreign investment signified turning back to the days of slavery. Patullo (1996, 10) noted that alongside Black Power calls for struggling against neocolonialism, the phrase "Tourism is Whorism" was also in play in recognition of the inherent racism that came along with the demeaning aspects of "recolonization." The increased dependency of tourism to keep the "islands from going down the tubes" resonated with foreign hoteliers and their allies, who saw the sector as a lifeline. By the 1980s, mass tourism was entrenched with more visitors traveling for sun, sea, and sand. Subsequently, as an economic, sociopolitical geographical location, the Caribbean became the most tourism-dependent region in the world. The tourist-sector contributions to the Caribbean region gross domestic product (GDP) was 14.8 percent by 2004 and by 2011 it was 16.6 percent (World Tourism Barometer Abstract Vol. 18 January 2020). This percentage varies by country while others are still economically bound to former colonial rulers, such as Martinique, an overseas department of France, Curacao, still tied to the Netherlands, and Puerto Rico, an unincorporated territory of the United States (World Travel and Tourism Counsel Research and Economic Impact; January 2020, p. 1).

In 1994, the 13.7 million stayovers (visitors staying over at least twenty-four hours) and cruise ship passengers coming into port was more than six-fold increase from twenty years before (Patullo 1996, 11). The most popular destination in the Caribbean that year was the Dominican Republic with 1.9 million stayovers, followed by the Bahamas with 1.6, while Jamaica saw almost 1 million visitors. Fourteen years later, the Dominican Republic remained the most popular tourist spot, with Cuba (tourism now an economic socialist strategy) coming in second followed by Puerto Rico. By 2019, the Dominican Republic was still the number one destination (6.6 million), followed by Cuba, Puerto Rico, and then Jamaica, showing 2.5 million stayovers (tourismanalytics.com).

Hurricanes and a series of global economic crises in the 1970s and 1990s were nothing compared to the downfall of the Caribbean due to COVID-19. First discovered in Wuhan, China, in 2019, the virus spread throughout the world, leading the World Health Organization (WHO) to declare it a global pandemic by February 2020. "CO" refers to the type of virus (corona), type VI, "D" the disease, and "19" for the year of discovery (www.cdc.gov). The impact of the COVID-19 pandemic saw worldwide travel plumet by 65 percent (UNWTO.org). For the tourist-dependent Caribbean, a 50 percent drop had catastrophic consequences for national economies and the livelihoods of women and men who work in the sector. Using end-of-the-year data (January–December 2020), the air arrivals/stopovers for the four top Caribbean destinations marked the following decline: The Dominican Republic −62.7 percent,

Cuba −74.6 percent, Puerto Rico −53.0 percent, and Jamaica −67.2 percent. Arrivals by U.S. citizens to Cuba had already dropped 21.9 percent due to 2019 travel prohibitions by the U.S. State Department. Looking back to March 2020, the Jamaican Government (GOJ) closed its borders in the face of the spreading of the pandemic of the novel coronavirus, COVID-19. This action brought the economy to a standstill. Without tourism, businesses stood to lose the equivalent of US$3 million a day in revenue that sustained 20 percent of the country's economy (Meade 2020). By June 2020, Jamaica reopened business for tourism when minister of tourism Edmund Bartlett announced a plan to establish a "resilience" or "tourism corridor" that ran from west to east along Jamaica's northern coast, where all travelers would remain. Vacationers had to remain in the tourism corridor and not be quarantined, or if outside of the tourist area, they had to enter fourteen days of confinement and their movements were closely monitored. Jamaica instituted a "travel authorization form" for entry into the country. Travelers from specific states, such as Florida and Texas and other hot spots in the United States, underwent further scrutiny as well as those coming from Britain and other parts of Europe. All of these restrictions remain in place as mutants of the virus still appear and access to vaccinations is not globally widespread. The UN World Tourism Organization (UNWTO) follows these trends as well as the Caribbean Tourist Organization, the Caribbean Hotel and Tourism Association, and government institutions such as the Jamaican Ministry of Tourism.

WORLD TOURISM ORGANIZATION

Set up in 1925 in the Netherlands, The World Tourism Organization began as the International Congress of Official Tourist Traffic Associations. Nine years later, the International Union of Official Tourist Propaganda Organization (IUOTOP) was created. It was renamed as the International Union of Official Travel Organization (IUOTO). After World War II the IUOTO, then housed in Geneva, Switzerland, recognized that tourism was now a part of the fabric of modern life. The demands of the tourist trade and the impact of those visitors on a society were causing governments to direct their attention to this economic sector in terms of infrastructure development and job creation. The intertwining of tourism and development established a specialized agency to keep abreast of issues at cross-international, regional, and national levels. In 1967, IUOTO members transformed their organization to address these common interests and to cooperate with other international bodies. A few years later the IUOTO's Extraordinary General Assembly adopted the statutes of the World Tourism Organization and installed the Secretariat in Madrid with the Spanish Government providing the initial infrastructure. Since 1979, the

UNWTO, now a special agency of the United Nations, has promoted technological transfers and international cooperation, stimulated and developed public–private sector partnerships, encouraged the implementation of the Global Code of Ethics for tourism, maximized the possible economical, social, cultural impact of tourism, and minimized its negative social, cultural, and environmental impact (tourismnotes.com). In 2003, there were 141 member countries, 7 territories (colonial appendages), and 360 affiliates representing educational institutions, tourism associations, and local tourism bodies (ibid.). One author characterized the UNWTO as the cheerleader of tourism.

The travel and tourism sector contributed nearly US$59 billion to the GDP in 2019. In that year, among the Caribbean islands, the Dominican Republic and Cuba showed the highest contributions from tourism to the GDP with US$14.3 and US$10.8 billion respectively. Since the 1990s, the total tourism contribution continuously increased bringing revenue, employment, as well as a list of social and economic woes, particularly to those nations who heavily rely on tourism for their survival.

A QUICK HISTORY OF JAMAICA AND TOURISM

In 1494, Spain became the first colonizer of the island that would be called Jamaica. This word *Jamaica* closely resembled the Taino name, *Xaymaca*, for the island, roughly translated as "the land of wood and water." The Taino (formerly known as Arawak) were the indigenous people and numerous landmarks on the island are named in that language. Nonetheless, at the time of the conquest, the Taino were in constant conflicts with the Spanish, leading to the genocide of these indigenous people. Although they claimed the island fifteen years earlier, the Spanish did not really occupy the island until 1509 and received little support from Madrid. These outposts in Jamaica were sparsely populated settlements that were just used as refueling stations. After 146 years of occupation, the Spanish were forced out by the British in 1655. As part of the expansion of the British Empire, and thanks to marauding Buccaneers, Brits began an almost 300-year rule of the island. Jamaica, the "land of wood and water," signified the island's bounty and natural beauty that turned into the green-gold of sugar cane, which the British thoroughly exploited. With the enslavement of African labor, fifty-seven sugar estates in 1673 grew to 430 by 1739 (Jamaica Tourist Board "Brief History" p. 1). After the emancipation of the enslaved, which began as an amelioration period in 1834 and ended in 1838, Jamaicans moved to where they could be free, take care of their families, and make a living.

Although British travel agent Thomas Cook promoted Jamaica as a place to "take in medicinal waters" for unhealthy Europeans in 1860, the island really entered the tourism industry in 1891 via the banana export business. Lorenzo Dow Baker of the United Fruit Company (now known as Chiquita Banana) often brought passengers with him on his weekly roundtrip sailing from Boston, Philadelphia, New York, or Baltimore to Jamaica. While he loaded up his ship with bananas, his guests took in the sights of this lush, tropical island. Within six years, Captain Baker acquired land and built Hotel Titchfield, an impressive 150-room edifice with elaborate dining facilities, and inaugurated the tourist business on the island. Advertisements from 1910 proclaimed Jamaica as "the most beautiful situation in the West Indies," and indeed it was a beautiful situation, as Jamaica offered a paradise of health seekers, winter tourists, and a new location for foreign capital investment (Taylor 1993, 45).

What this first tourist hotel produced was a new site for interpersonal service between peoples of different social classes and in this particular case, different skin color. Just sixty years out of slavery, but still a British colony embedded by its hierarchically arranged social system based on color, class, and gender, Jamaicans reluctantly accepted tourism as another sector of the economy. The early hotel industry served to resuscitate the dying master–servant culture of the Great House era in Jamaica, particularly from the perspective of wealthy White clients. In fact, early tourism in Jamaica did in fact resemble slavery, particularly when Jamaican hotels were reserved for White-only Americans, while Black Jamaicans were left with the menial tasks (Patullo 1996, 64) and were certainly not guests. By the late nineteenth century, Jamaica emerged as an escape for the wealthy and weary from the metropoles of Europe and America, with the emphasis on U.S. and Canadian visitors. Frank Taylor (1993, 93) wrote, "Tourism had come to represent a new kind of cash-crop—a new sugar."

The tourist industry faltered during the years of World War I, but it came to life with a focus not on the medicinal waters and the mountains but on the sea and sunbathing. Even during the Great Depression years, the growth of tourism benefited from a successful publicity campaign in Britain supported by some steamship companies and travel agencies that extolled Brits to see the island, following the sentiment of "seeing the Empire first" (Taylor 1993, 144). For example, in 1937 the Jamaica Tourist Development Board recorded 65,690 visitors (10,432 stopover and 54,837 cruise tourists) to the island. At that time, Jamaica benefited from political turmoil in nearby Cuba, Italian fascism, and its conflicts with Ethiopia, which made the Mediterranean totally uninviting. Subsequently, vacationers diverted their travels to Jamaica. Coincidently, in 1930 Pan American Airlines first linked Jamaica to the rest of the world when it inaugurated its service through Kingston on its Miami–Panama route. According to Taylor (1993, 156), the first flight carried twenty passengers per trip from Kingston to Miami in about six hours.

By the end of World War II, the history of Jamaica's tourism rested on the available and lavish service for middle-class vacationers who did not have this kind of attendance at home and were still recovering from the war. In 1947, Cunard Steamship Line, one of the prestigious modes of travel at that time, featured Jamaica as a premier travel destination for the wealthy. Montego Bay also had an additional advantage of a good-sized air strip that was laid out during World War II for allied defense forces. By the 1960s, the regular long-haul air services from Europe and the United States ushered in a new era in the tourist industry and a new kind of tourist. Frequent airline travel opened up the island to mass tourism whereby the middle-class traveler from the United States could and did demand service for his or her hard-earned dollar. This period is considered the gilded age of Jamaican tourism, as Montego Bay on the North Coast was literally its gold coast.

Due to the island's proximity to the United States, Jamaica, and Mo Bay in particular, saw its number of visitors rise way above those of the other British West Indian islands. Between 1945 and 1962 the number of hotels on the island increased twofold (Taylor 1993, 162). The industry shifted its time-table, adapting from catering to the upper classes (December to April) who had the means to holiday during the winter months, to the summer months when the prices for accommodations dropped in price, making it more afford-able for mass tourism customers.

During the early days of its independence, the Government of Jamaica expanded the role of the Jamaican Tourist Board (JTB), the national tourism agency. Founded in 1955, the JTB and other related government agencies set to further promote tourism, which was increasing in significance in the econ-omy. Hoteliers were given the status of "duty free" to import building materi-als, furniture, equipment, and other supplies in efforts to expand the capacity of the tourist trade. This government package also included the proviso of a ten-year tax holiday for investors. There was the Hotel Incentives Act of 1968, the Resort Cottages Incentive Act 1971, and the Attractions Incentive Regulation, all playing major roles in the development of the sector. In 1968, Air Jamaica the national airline was launched and several airlines flew into the two expanded airports—Sangster International Airport in Montego Bay and Norman Manley International Airport in Kingston.

By the 1970s, the affluent, wealthy mobile traveler continued to visit Jamaica, but the majority of guests were middle-class Americans looking for bargains as well as unaccustomed luxury. However, a mid-decade conflu-ence of global events changed the course of the society. First, the price of crude oil increased due to the 1974 oil embargo. Revenues from Jamaica's bauxite mining and failing sugar production could not keep up with the need for foreign exchange that precipitated financial duress. The country entered a debit crisis. Second was the destabilization of Prime Minister Michael

Manley's 1970s government. His political ideology promoted self-reliance in all aspects of citizenry and questioned the motives of the United States in its economic relations with Jamaica. Manley's leadership in the nonaligned movement and friendship with Cuba raised a "red flag" in the international banking and trade community that eventually led to the country's first negotiations with the International Monetary Fund (IMF).

Third, in the tourist sector, Jamaica pursued at least three marketing strategies to assuage visitors. Sensing a "red scare," and part of the power of global market forces, tourist agents and their middle-class, mass-market clients looked elsewhere for their vacation spot. Adding to the problem, many of the international chains such as Hilton and Holiday Inn left the island. Hence, the first JTB plan of action promoted Jamaica as "more than a beach but a country" and provided alternatives to mass-market tourism targeting the adventuresome overseas visitor. The second tourist plan focused on domestic vacationers as the government and its tourist agencies aggressively encouraged middle-class Jamaicans to vacation at home in Jamaica, rather than travel to the mother country of the United Kingdom or to the United States. Jamaican companies were offered tax packages and incentives for their staff to discover their own island, hotels, beaches, and other accommodations. In addition, Jamaicans living abroad were encouraged to add a hotel vacation segment to their plans when they came home to visit family. The Jamaica Tourist Board had to sell Jamaica to Jamaicans. Furthermore, Jamaicans were participating in the industry as guests but also as entrepreneurs. During this moment of self-reliance, the Government of Jamaica helped citizens to finance their ownership of hotels. A stipulation was that new owners also had to learn hotel management and to develop training sessions. As a result of these initiatives, when the third strategy came on stream, hotel owners, managers, and employees were ready. In 1976, the "all-inclusive" resort Negril Beach Village, owned by Jamaican hotelier John Issa, became the prototype of what was to become one of the most revolutionary products of tourism worldwide. Details of "all-inclusive" resorts are examined later on in the book as a source of employment.

Following the defeat of Michael Manley in the 1980 election, the Jamaican Labour Party (JLP), under the leadership of Edward Seaga, inaugurated a new period of conservatism that ushered in pro-business, free-market capitalism schemes into the economy. This economic direction followed the path of structural adjustment policies (SAPs) designated by World Bank and the IMF that fit into the JLP platform of denationalization and deregulation. The country found itself involved in three new financial and developmental areas: the introduction of more SAPs; massive loans from the World Bank, Western commercial banks, and a variety of international lenders; and U.S. investors in industry, whose financing was facilitated by provisions of the U.S. trade agreements promoted under the Caribbean Basin Initiatives I and II (Bolles,1996). The gains made in Jamaica exacted a high price from Jamaicans themselves as violence and hardship, due in part to the deterioration of social service systems, escalated (see Davis and Anderson 1987 for the human impact of these policies).

Jamaica's tourist sector was a most hospitable arena for free-market investment. Early on in the 1980s, the JLP government divested numerous hotels purchased by the previous PNP administrations. The newly established Ministry of Tourism proclaimed the government was seriously committed to tourism and by 1983 the sector was the largest earner of foreign exchange for the Jamaican economy (Planning Institute of Jamaica Economic and Social Survey Jamaica 2001).

Jamaica's share of the Caribbean tourist market also grew through the expansion of cruise ship arrivals to Montego Bay and Ocho Rios, with stopover visitors numbering over half a million in 1985. Seventy-five percent of vacationers coming from the United States found an additional attraction besides sea, sand, and sun and affordable pre-paid package deals. Although many vacation packages included meals and accommodations, activities outside of the resort were not included in those deals. Those events were paid in local currency, a bonus for the tourist. One of the outcomes of Jamaica's adherence to SAPs was the devaluation of its currency, which favored tourism. For example, in 1985 a tourist seeking a two-hour ride on an off-site glass-bottom boat paid JM$60 (5.58 = 1US$), which cost that visitor roughly less than US$33. With gasoline priced at US$2.19 per gallon, there was scant profit for the boat owner.

When the Peoples National Party returned to power in 1989 with a different political philosophy, the tourist sector was flourishing despite infrastructure setbacks caused by the devastation of Hurricane Gilbert in 1988. Working together in a joint effort, government and the private sector poured capital, incentives, and expertise into the sector. In 2001, filmmaker Stephanie Black directed a documentary *Life and Debt*. The film examined the policies of the IMF, World Bank, and other international lending agencies that had changed the economy of Jamaica over the course of twenty-five years. The film juxtaposed wealthy Americans vacationing in luxurious resorts safely isolated from Jamaicans living outside of the resorts' compound who struggled to make a living under dire conditions of a devalued currency. The IMF guidelines gutted national expenditures for social services such as health care and education. Interviewed in this film were construction workers on a resort site, dairy farmers who lost their farms due to competition with foreign products, and cultural and academic critics.

Nonetheless, critical to this neoliberal momentum was the money that Jamaicans were investing in the sector. Even pension fund managers invested significant sums into the sector. By 2002, 80 percent of Jamaica's tourist industry was owned by Jamaicans (Bolles 2008). The island's natural beauty, strategic aggressive marketing by the Tourist Board, and new local captains of industry made Jamaica one of the leaders of global tourism.

Two Jamaicans, John Issa (Super Clubs and Breezes) and Gordon (Butch) Stewart (Sandals and Beaches), broke into the American and European lock on all-inclusive holidays. Tourist researchers called this business endeavor "the most important innovation in the Caribbean hotel sector during the last decade" (Issa and Jayawardena 2005, 223). Both Super Clubs and Sandals were homegrown marketing successes and copied all over the region. Stewart claimed that the success was based on value for money. "We have the biggest water sports business and fitness centers, brand-new restaurants, great entertainment. You have quality choices and with all that you end up with value for money you can't get anywhere else in the world" (Pattulo 1996, 21). Originally an owner of a car parts dealership and appliance store, Butch Stewart started Sandals in 1981 by remodeling an old hotel, turning it into an all-inclusive "couples only" resort (see preface). Three decades later, the Sandals chain, which also included properties called "Beaches," owned and operated sixteen hotels, seven resorts, including five deemed "luxury" in Jamaica, as well as others located in St Lucia, Antigua, the Bahamas, Barbados, Turks and Caicos, and the Dominican Republic. In 1992, Stewart personally diverted a national foreign exchange crisis, when he deposited US$1 million into the Jamaica's commercial banks for less than the exchange rate. This effort kept the banks and the foreign exchange rate from collapsing. It also encouraged other Jamaicans to put their U.S. dollars into local banks, thereby stabilizing the local currency. Following this action was another coup for this entrepreneur. Butch Stewart relieved the government of its controlling percentage of the national airline, Air Jamaica. Needless to say, Air Jamaica flew to all of the countries where a Sandals resort was located, basically making his corporation a totally integrated operation. By 2011, however, Stewart divested his holdings and after years of debt sold Air Jamaica, which is now merged with Caribbean Airlines. Never skipping a beat, the Sandals Corporation continuously wins countless awards from the hotel and tourist industry. The Issa family developed one of the first properties for mass tourism in the 1950s, Tower Island, and by the 1980s–1990s the Issa corporate holdings were found throughout the region. In 2009 the Super Club brand was dismantled and a number of properties were sold. In Negril, the re-branded and renamed resorts were Breezes Resorts and Spas, Hedonism Resorts, and Rooms on the Beach.

LEAKAGES AND ISSUES

Over the years, a major concern in tourism focused on the value of the contribution of the sector to the local economy. In 2012, a report commissioned by the Jamaican Hotel and Tourist Association (JHTA) assessed the benefits and leakages of tourism. Listed attributes included the following:

1. Tourism's linkages to the rest of the economy were strong, extending its total economic impact to 19.5 percent of GDP.
2. Tourism is Jamaica's #1 export sector.
3. Tourism has posted consistent growth, even when the rest of the region was in decline.
4. Tourism contributes substantial tax revenue and further taxes would jeopardize its competitiveness as an export industry.

Ongoing issues and problems included the following:

1. The hotel sector has a relatively high tax burden in comparison to regional sectors.
2. The incentives offered to the industry have been an essential part of maintaining growth while they remain generally less generous than those offered across the Caribbean.
3. When factoring in linkages, the tax contribution of tourism is greater than any other industry and greater than its contribution to GDP.

Nonetheless, the leakages, the value of the contribution of the sector to the local economy, and the high price of imports of food and consumer goods for both locals and tourists remain ongoing problems (TE Oxford Travel and Tourism 2019).

As the above list showed, tourism, a broadly defined service-producing sector, has been the single largest generator of foreign exchange for Jamaica. In 2018, 34 percent of the country's GDP, with earnings of US$5.2 billion, came from tourism and travel. For the following year, there was an increase of 34.7 percent with US$2.3 billion.

THE SPANISH INVASION

For years Sandals Resorts and other locally owned and operated hotels dominated Jamaica. But since the mid-2000s as Jamaican sociologist Orlando Patterson (2019) and other scholars noted, more Spanish-operated all-inclusive resorts were coming on stream than ever before. RIU and other conglomerates are heavily invested in tourism in Jamaica. The repercussion of these non-Jamaican corporations was felt on two levels. First, the major chains, such as Sandals, welcomed the competition and developed new products to entice a more upscale clientele than those attracted to the budget mass tourist group that flock to RIU. In an interview with Joe Pike (2009), Butch Stewart said that, whereas the Spanish all-inclusive resorts like RIU provided much more rooms than Sandals did, they cannot compete with Sandals' personal service. In fact, Stewart noted that Sandals had two staff members per room while RIU and other all-inclusive like it have about .5 staff members per room. The underlying difference, as Stewart argued, is the essence of good and plentiful service and the quality of receiving guests housed in a sense of professionalism that is the accoutrement of hospitality in tourism.

The second impact of RIU and other Spanish-owned properties in Negril and elsewhere on the north coast of the island is the disregard of the local environment and building codes. Filmmaker Esther Figueroa's documentary film *Jamaica for Sale* (2006) presents the case of economic and environmental problems facing the country. Implied are acts of bribery and intimation of incompetent and corrupt public officials. Interviewed workers report of toxicity of waste runoffs, their poor pay, and hazardous working conditions all the while demolishing the natural coastline as they construct a Spanish-owned resort. The film "counters the dominant view that tourism is the savior of the Jamaican people that is lively and hard hitting with powerful voices, arresting visuals and iconic music that shows the cultural impacts of unstainable tourism development" (comments from the Caribbean Studies Association meetings, Kingston, JA, 2009, where the film was launched). The privatization of coastal lands leaves Jamaicans without access to beaches, reduces the local population to service jobs, and makes owning land on their home island impossible for most Jamaicans ("News and commentaries" Repeating Islands, 2009, p. 1). In an interview, Dr. Figueroa said that the title of the film was flexible as people from all over the world told her that their country too was for sale to foreign investment.

BRANDING OF JAMAICA

No matter the political party in office, tourism is the "cash cow" for the country, with significant trickle-down flows for those in the middle that can even reach those on the bottom of this sector. Fueled by advertising since the later part of the nineteenth century, from the 1980s when the lure was to "return to how things used to be" or "to feel good all over," tourists are most welcomed in Jamaica. In the ensuing years, similar sentiments have come from the Jamaica Tourist Board, such as the 1994–2003 campaign of "One Love . . ." followed in 2003–2007 with "Once You Go . . . You Know," and then in 2012–2014 came "Home of All Right," and in 2016 "Join Me in Jamaica" (www.visitjamaica.com). In January 2020, the "Heartbeat of the World" slogan was launched, reinstating Jamaica as a global leader among travel destinations. Thus, the "Heartbeat" ad campaign established Jamaica as the single destination that every traveler must experience (Jamaican Tourist Board "Join me in Jamaica" campaign, January 8, 2020). Further, Donovan White, director of tourism for Jamaica, remarked,

> On the map, Jamaica may seem like a small dot in the Caribbean Sea. But Jamaica's influence on the world culture is the size of a continent. We are a cultural giant, and we continue to have an indelible impact on the world's food, music, sport and literature while offering larger-than-life bucket list experiences with extraordinary, talented people. (ibid.)

Promotion of all things Jamaican has a place in the history of selling its tourist offerings.

POSTCARDS

Postcards, now eclipsed by email and photo sharing on digital cameras, were essential tools of "branding" for Jamaica. One of the critical elements of branding is promoting a product, symbol, or a term that is a full expression of or perception of how that item is reputed to be. Here is a discussion of three postcards that are illustrative of how the reader/audience of that time was "sold" on this branding of Jamaica. The 1940s postcard was one of the items that lured post–World War II tourists to Jamaica. Postcard images featured something exotic, different, and unfamiliar all the while conjuring up the idealized adventure. Idealized was the Jamaican peasant woman with a basket of fruit on her head standing beside a donkey with the greenery of palm trees and bright pink- and coral-colored floral blooms in the background. The visages were particularly true for the middle-class European/White North American who beforehand could not travel for pleasure. Further, these kinds of images served perhaps for the viewer's/potential tourist's own edification without their traveling abroad for the classic trip to continental Europe.

By 1962, following Jamaica's independence, postcards continued featuring the same tropes that idealized the country's exoticism for tourist consumption. A popular postcard showed four different images that represented what was in store for the visitor. As usual, depicted was a smiling Black peasant woman with a basket of fruit and vegetables on her head. The other images were efforts to illustrate the "modern" Jamaica where self-identification was part of the promotional vehicle. One could imagine that during a trip to Jamaica the viewer could consider playing golf, enjoying water sports, and swimming in calm aquamarine water. The photos in this card are framed by tropical fauna, lots of palm trees with additional greenery provided by the mountains as a backdrop. Alongside that image was another postcard of the mid-1970s that featured the five national symbols of the country: the National flag of black, green, and gold, the Jamaican coat of arms, the national fruit (the Ackee), the national flower (Lignum Vitae), and the national tree (Blue Mahoe). These colorful images conveyed the modernity of Jamaica as an independent country that prided itself on its historical past, which was part of the "we are more than a beach we are a country" JTB campaign. Nonetheless, the perspective of the potential visitor was critical in the postcards circa late 1970s/1980s reinforced by television advertising and tourist agency promotional materials. All of these images were used to illustrate the adventures big and small that awaited the visitor.

THE POSTER

In 1972, the JTB contracted the innovative and prestigious advertising giant Doyle, Dane, and Birnbach to create a campaign for the growing tourist sector. This advertising company was recognized for its lively and entertaining copy, and a photographer was sent to Kingston to scout out locations and subsequently caught the "image" of Jamaica. There are two print ads deemed exceptional by the advertising world from this campaign. According to communication arts, "DDB's printing ads for Jamaica was the most beautiful and probably the most informative travel advertising ever done." Germane to this discussion is the "Tee-Shirt" poster.

Beforehand, the image of Jamaican women was viewed as symbols of the peasantry, holding a basket of fruit on their heads, with the smiling face of welcome to visitors. In this ad, the beauty is not only of the landscapes of beaches, the fauna, flora, and the like but also of the lushness of young Jamaican women. The 1972 T-Shirt poster reimagined what modern-day young Jamaican women looked like. Is it this beauty that brings vacationers to Jamaica, or does this ad, which also appeared predominately as a poster, go alongside other "pin-ups"? Is this ad something to gaze upon, but never to hold up closely in one's personal space? The subtext has heteronormative, exploitative, racist, and sexist overtones.

The poster graced the walls of airports, JTB offices worldwide, hotel lobbies, board rooms, travel agencies, and museums internationally. The orange-red wet T-shirt clings to her body, caressing her full bosom as it extenuates her perfect "10" figure. In Jamaica, there was a bit of mystery as to the identity of this woman. In 2000, it was finally revealed that Trinidadian Sintra Arunte-Bronte, who was visiting Jamaica at that time, is the JTB poster "girl." The photographer from Doyle, Dane, and Bernbach, looking for a woman who could portray the beauty of the island, found the young woman, who agreed to be photographed. She stood outside of the Pegasus, now Le Meridien Jamaica Pegasus in Kingston, for seven hours for this poster shoot. According to local lore and the Jamaican Tourist Board, Sintra Arunte-Bronte received no more than a few dollars and never received any royalties for the poster that is now a collectors' item (Annica Edwards, p. 1, Jamaicans.com). In a blog for Lonely Planet (February 14, 2010), Sintra Arunte-Bronte says that because of the popularity of the advertisement, people recognize her as the poster girl and ask her for her autograph.

TOURISM: ITS PAST AND IMPACT

Throughout the history of European expansionism, Caribbean tourism was connected via the leisure travel of the nobility and the upwardly mobile upper class. These groups of travelers ventured out to the tropical outposts for health and adventure. The innovations of the industrial revolution and entrepreneurs such as Thomas Cook made it possible for the middle classes to join in on the fun. Cook made it easy for this new set of travelers to become well-acquainted with the technology of the day and encouraged others to partake in this adventure to the tropics. Beginning with thousands, now millions of visitors to the Caribbean, the region is dependent on the United States, Canadian Euros, and British currencies for their economic well-being, As George Gmelch (2012, 12) quoted a promotional bit of literature, "The sun never sets on the tourist empire." Tourism is the new sugar.

Due to the inequitable nature of tourism itself, now part of tourist lore is the case of the arrogant worker. These stories are readily available in internet chat rooms for perspective clients to read and perhaps take their business elsewhere. Twenty years ago, a popular tourist guide (Baker 2000, 66) said this, "Jamaicans are an intriguing contrast. Most of the population comprises the most gracious people you'll ever want to meet. [However] a significant minority are composed are the most sullen, cantankerous, and confrontational people you would ever not to meet." Foreign visitors are often shocked at the surliness they so often encounter. The thought that every Jamaican should be hospitable to all who visit their country—particularly if they are Euro American, Euro-Canadian or European—speaks of how unlearned visitors can be about Jamaica.

Most travel to Jamaica is aided by PR tourist imagery of sun, sand, and sea and not for history or sociological lessons. The popularity of all-inclusive vacationing shields visitors from the problems that Jamaicans face on a daily basis. The price of food, consumer goods, clothing, shelter, poor roads, and other infrastructure issues; crime; drugs; lack of access to adequate health care; and lack of quality education are challenges for the majority of

Jamaicans. Many of those services are shortchanged in order to satisfy the national debt to international lending agencies that financed the building up of the tourist sector. Tourists, who are unaware of these issues, make for the inevitable rancor when there is an encounter. Fortunately, the popularity of heritage tourism in Jamaica might prove the sociocultural panacea where there are travelers who do look for history, culture, and sustainable development causes when they come to Jamaica (Francis-Lindsay 2009). Unfortunately, heritage tourism is not on the agenda here. However, the "Capital of Casual," the name promoted in Negril, has something else to offer for adventuresome travelers, as discussed in the following chapters.

Chapter 2

Tourism in Negril, the Capital of Casual

A Rastafarian artisan from Trelawny had this to say about tourism:

> Hotels are not the main reason why people come to Jamaica. Good hotels are all over the world. What makes the tourist industry special to visitors is the hospitality of the Jamaican people, and the fresh unique, natural products, music, culture and other services offered by Jamaicans. (Dunn and Dunn 2002, 27)

This sentiment echoed the lure of Negril as a tourist destination. As *Negril Guide 2020* (Negril Chamber of Commerce) touts, Negril, this "little piece of paradise" Negril is the Capital of Casual founded on interpersonal connections. This appeal is based on that unique, natural style that harks back to its origins as a small fishing village. Even if a contemporary visitor does not venture out of the security of an all-inclusive, the attraction to Negril stems from the of white sand beach, laid-back atmosphere, breath-taking sunsets, reggae, dancehall, and other cultural events. Although the words of that Rastaman from Trelawney were uttered just before the new millenium, they form general directions for tourists who venture to this most western part of the island. It is estimated that the tourist sector directly and indirectly employs 90 percent of the residents of Negril and still attracts workers from nearby towns and villages who seek employment in that locale. This is just one side of this many-sided tourism coin.

A history of tourism in Negril provides the physical and research setting of *Women and Tourist Work.* It begins with a brief accounting of the place and how it got its name. Included is a discussion of the pre-1970s economy. Much of the early history is attributed to a 1950s anthropological study and followed up by the work of other anthropologists in the late 1970s. The next section examines the 1970s growth and development investments through the Government of Jamaica's Urban Development Corporation (UDC). The discussion then addresses the social and economic impact of tourism in Negril and illustrates the strengths and weaknesses of tourist dependency. Further, given the historical rural character of Negril, the role of small business is voiced by the Negril Chamber of Commerce. As a member organization, the NCC has become the caretaker of the village with its community orientation, which serves as a bridge between guests and hosts for this village, now a tourist mecca.

THE STORY OF NEGRIL

Once considered a swamp of mangrove trees, Negril was a sparsely inhabited coastal community situated on Jamaica's remote west coast separated from the mainland by the Great Morass. Its name might have derived from the Spanish *Anguila Negra*, associated with the bounty of little black eels found in the local waters. Or, according to a tourist guide (Visit Negril Guide 2020, p. 10), the name *Negril* is a shortened version of the Spanish *Negrillo*, referencing to the dark-colored cliffs that lead to the western part of the area. No matter the derivation, over time the name was anglicized and shortened to *Negril*. In 1915, G. Logan McLeod purchased 1,800 acres of land in Negril from the Santiflebens of Lucea. The rest of the beach belonged to local fishermen, who then split these properties among their family units (*Negril Guide 2020*, 10). Although some were still wage laborers on the nearby Frome sugar estate in Westmorland, this close-knit group of families made their living by fishing; they owned their canoes and sold fish in the markets of Savannah-La Mar and Lucea. Women tapped the native Jamaica palm coconut trees, located on the water's edge, sold its oil for the market, and tended to garden plots that yielded food for the nourishment of their own families. Featured in *Negril Guide 2020* is the story of Rhoda Jackson, who was a member of a well-known family and raised four children, who then sired sixteen grandchildren. This brief vignette tells of the times before Negril and tourism. Rhoda Jackson's weekly routine required her to go to the market to sell the coconut oil produced by her family. She walked the fifteen or so miles from Negril to Savannah-la-Mar with containers filled with coconut oil balanced on her head. She did this trek for decades. Fifteen miles going in the other direction along the coast from Negril to neighboring Green Island was also done on foot or by the sea. In 1944, Ivan McLeod with local fishermen hacked out an eight-foot-wide, five-mile-long sandy-dirt road between those two settlements (ibid.).

In the late 1940s, a disease swept through the prevailing winds, killing the tall coconut palms. This agricultural calamity slowly eroded the livelihoods of those who worked on all levels of production, from harvesting the fruit, to extracting and processing the oil for household consumption as well as for profitable distribution to Savannah-la-Mar markets. An example of this livelihood comes from the recounting of Rhoda Jackson's story whose family owned a home on the beach and produced coconut oil. Without the trees, the result was a wide sandy beach that shimmered in the sun.

NEGRIL OVER TIME

Over the course of four months across two summers (1954 and 1955), Yale graduate student of anthropology William Davenport conducted his dissertation research. It was a community study of two fishing villages in Jamaica, one of them being Negril. He assessed whether people—mainly the men— were full-time fisherman or if they supplemented their seafaring with being part-time cultivators or full-time cash crop agriculturalists. He described the village as consisting of three areas: the West End, Red Ground, and the beach. At that time there were 1,161 houses made of differing construction materials and building types that ranged from hovels with thatched rooms to multiple-room wooden structures. Davenport noted that at the back of each house was a cleanly swept yard with a cookhouse/"kitchen." People viewed their own history as starting with Emancipation (1838), and all who participated in Davenports' work noted that they always lived in Negril. The village was serviced by the Royal Mail bus, which left the post office late every afternoon except Sundays for Savanna-la Mar (15 miles away) and was met by another bus carrying the incoming mail to Negril, making the return trip around midnight. Davenport recorded that on Saturday morning a truck went from Little London (off the West End Road) to pick up women along the way who were going to market in Savanna-la-Ma. The truck would wait for the women to bring them back in the evening. In the 1950s, this trip took hours. People walked or rode bicycles, mules, donkeys, and horses when they traveled in and out of the village. Among the institutions were a primary school and an Anglican Church.

As the tall palm trees toppled and the oil extraction came to an end, the denuded beach front provided the possibility for a new economic opportunity exemplified by the role that Rhoda Jackson's family played in contemporary Negril. Thus began the scramble for acreage following the lead of the tourist business in the North Coast Montego Bay area. Davenport (1956, 56) wrote, "Negril has more undeveloped tourist potentialities than other beach areas on the entire island . . . and the search for beach house sites is very attractive again for Negril." Negril was poised for change.

Although identified by government officials as a most attractive site because of the seven miles of white sand beach, Negril was late in coming into the world of tourism. This was a slow transformation from a peasant coconut-processing area and fishing hamlet to one geared for tourism, as prophesized by Bill Davenport in his 1956 dissertation. A reason for this almost leisurely pace had to do with Negril itself, still a fishing village and home of Rastafarian artisans and farmers. Already in place were small family lodgings that welcomed visitors who took advantage of this relaxing, unhurried Negrillian atmosphere. In the early 1960s, middle-class Jamaicans still got away from it all in Kingston and rented cottages and rooms from local people in Negril. Moreover, as a seemingly isolated village, there were infrastructure issues to be addressed before anything could transpire. Negril and its residents took time for this development to occur, and the government was slow to fund the projects. For example, the dirt/sand rock road, also known as the North Coast highway, that connected Negril Centre to other coastal thoroughfares that reached Montego Bay remained the roadway until 1959 when construction finally started. Other infrastructure undertakings began at that time such as extending the main road to Sheffield, canalizing an extensive portion of the Great Morass to facilitate increased drainage and make land available for agricultural use, and constructing a water treatment plant. All of these tasks were completed by 1961. Later on, electricity and telephone service arrived in Negril (Rhiney 2012, 26). All of these public works were accomplished to make Negril accessible for tourism, which was viewed as a postwar development strategy (ibid.).

There were three additional problems. One was that even after construction of the North Coast highway from Montego Bay to Negril, it was still a winding and narrow thoroughfare and often filled with potholes. Second, land, especially beachfront property, was being sold by the ensuing land speculation by potential investors. The third was the indecisiveness of the government as to what kind of development strategy was most appropriate for the area (Rhiney 2012, 30).

THE TRANSITION PERIOD

Recounting his first trip to Negril, former prime minister Edward Seaga (2006, xxiii–xxxix) wrote:

> I first visited the area in the late 1950s. I came upon it by chance, simply through exploring Jamaica. Driving through Negril on the main I got a glimpse of the sea, here and there. I was curious about the possibility of an extensive beach front as the sea kept appearing mile after mile in little windows of opportunity

sightseeing. Finding one opening which allowed passage to the sea, I drove in and parked. I was stunned by the miles of uninterrupted beach that I saw. It was awesome. As I walked along the beach, I came upon a series of little houses every 30 yards or so, set back from the sea. I struck up a conversation with the first person I saw. I learned that this was called Long Bay and the beach was seven miles long. I also learned that generations of families had lived on the long narrow plots of land, half of which were on the beach side of the main road. The other half were on the other side, which fronted on a huge pond called the Great Morass. . . . What I saw left me in shock: another beach about two miles long, shaped in the arch of a bay. . . . The water was calm and azure blue, the sand white and crystalline.

Mr. Seaga went on to say that in 1968 when he returned to Negril on vacation with his wife, he saw the possibilities for the development of a mega resort. By this time, Jamaica was an independent country, and Seaga was a leader in the Jamaica Labor Party (JLP), which won the first election, and was a Member of Parliament (MP) representing West Kingston. He reminisced, "The Great Morass with great care to protect the ecology could be a prime attraction, and medium-sized hotels would fit into Long Bay with smaller establishments on the West End Cliffs" (p. xxv). Now, he remarked, "I heard that Ivan McLeod was the owner of the properties and asked him if he would sell the land to the Government" (ibid.). McLeod was receptive, and after a discussion about price and valuation, "the deal was closed." One of Mr. Seaga's accomplishments as an MP was to establish the UDC, which invested in a Land Bank focusing on waterfront development in coastal towns and rural areas. Negril was put through the planning process. Seaga noted, "The first hotel was called Hedonism to reflect the aura of closeness of nature and hedonistic pleasure in this paradise" (ibid.).

In the aftermath of Jamaica's independence in 1962, U.S., Canadian, and European hippies who usually camped on the beach in tents and young professionals on a tight budget came to this relatively secluded western part of the island. This was the basis of the word-of-mouth promotion of the laid-back lifestyle for which Negril was well-known. In his semi-autographical novel *Banana Shout*, Mark Conklin (2000) recounted the transformation of this beautifully isolated village of fisherman and Rastafari that slowly encountered this invasion of hippies, Viet Nam draft dodgers, and drug smugglers. Conklin wrote (as his character arrived via boat),

As they drew closer to Jamaica, the water changed from deep navy blue to crystal-clear emerald green and the distant mountaintops loomed lushly with tropical rain forest splendor. Then to his delight, and astonishment there appeared the most pristine, white-sand beach he'd ever imagined, stretching endlessly for miles, surrounded by a teaming coral reef. . . . He felt he was having a religious experience. (53)

Later on in the novel, a character contemplated, "For him Negril was a simple peaceful fishing village populated by honest, hardworking, sincere country folk who lived off a land and sea unspoiled contamination" (129). However, another character in the novel had other visions in mind. Nick screamed, "You fool! We'll take paradise and put in a parking lot" (129).

Tourism development began with large-scale construction of several resorts in Negril that was emblematic of the sector in the country. This development was driven by a mix of state incentives such as tax cuts to import duty concessions. By the 1990s a total of five Hotels Incentive Acts (1968, 1971, 1972, 1985, and 1991) passed by whatever political party was in power, allowing for ten years for approved hotel enterprises and up to fifteen years for convention-type hotels. Rhiney (2012, 27) noted that Negril benefited both directly and indirectly from these various incentives and partnerships. Negril Beach Village (renamed Hedonism II), the first Jamaican Government (GOJ)–owned hotel, in 1976 was subcontracted to a private-run resort management of Super Clubs headed by John Issa. The Lucea-Negril water supply system was expanded in 1989, cofinanced by the GOJ in collaboration with the Japanese government and USAID. All of this was to benefit the expansion of tourism in Negril. GOJ funding was allocated through access to loans from international lending agencies, thereby increasing the national debt through neoliberal economy largess. As noted earlier, in 1968 the conservative pro-business Jamaican Labor Party was instrumental in the founding of the UDC. The UDC continued to be the conduit to transform urban centers and rural towns while preserving the natural environment, traditions, and customs, and spurring economic development. It also provided opportunities for private capital stakeholders to participate in these endeavors.

In the late 1970s, anthropologist Deborah D'Amico-Samuels (1986) conducted research examining the impact of tourist development in the lives of the people of Negril. Her work combined careful ethnographic details that were framed by a political, economic, and gender context. Her work provided the insights of community involvement in the ongoing process of Negril becoming a tourist destination. She demonstrated how Negrillians used government infrastructure improvements to their own benefit and secured their inclusion in the growth of tourism in Negril. D'Amico-Samuels remarked,

> In contrast to the government plans which included a role for Negrillians only as laborers . . . tourism in today's (1980) Negril provides many residents an income without working for someone else, thus realizing the dream of many Jamaicans to "not have anyone over me telling me what to do." (112)

Enterprising Negrillians constructed cottages to rent to visitors as an alternative to the large and foreign-owned hotels. Although the simple concrete guest houses built by the community people were scorned as "ugly little boxes" by tourist officials, in the eyes of Negrillians and their admiring neighbors, the proof of development lay in the quantity and quality of these durable concrete buildings. Equipped with electric lighting and running water, they were visible signs of an improved way of life. They were a hint at another important characteristic of tourism development in Negril, its integration with other income-earning strategies. The fruitful trees and gardens that surround Negril's guest houses afforded a source of food as well as admiration by foreigners. Thus, at the time of D'Amico-Samuels's research, virtually all Negrillians who owned guest houses still practiced cultivation and fishing. A UDC checklist of facilities for 1980 showed the continued investment by 150 percent in fishing boats in Negril. Unfortunately, within a decade the profusion of even modest-sized fishing vessels led to the depletion of fish in the waters surrounding Negril.

It is no surprise that the UDC played a pivotal role in the development of tourism in Negril enabled by an increased national debt. The GOJ debt to the GDP was 126.57 percent from 1980 to 2017. The debt reached its all-time high of 212.40 percent in 1984 (tradingeconomics - Jamaica) when tourism development was in full swing. In 2018, Jamaica had the fourth-highest national debt of 94.4 percent within the Latin America and Caribbean region, according to Statista.

By the turn of the twenty-first century, Negril had increased hotel occupancy by 62.1 percent with fifty-three hotels in operation. This figure increased by 77.0 percent by 2018 with sixty-eight hotels (UNWTO 2019). In 2019, the vast majority of visitors stayed in hotels, while private homes, resort villas, guest houses, and apartments providing lodging together totaled for 467,671 tourists (UNWTO, ibid.). Although the counter-culture group of the early 1970s and Jamaican vacationers looking to "cool out" could be counted on, Negril still attracted travelers looking for less luxury, less structure, and more adventure for their stay in Jamaica. Nonetheless, Negril morphed into a resort location now open to all, but with a range of accommodations. There were now car ports and parking lots nestled in and crowding out space on the beach.

ON THE ROAD TO NEGRIL TO AN
ALL-INCLUSIVE VACATION

En route from Sangsters International Airport in Montego Bay, one passes the large-sized seaside resorts built along Route A-1, the new and improved North Coast highway that goes from Montego Bay to Negril. A traveler can glimpse past houses of local people that might block the view, but the glistening aquamarine sea and seafaring activities peek through the visage. The beauty and lushness of all fauna and flora come into view, with bananas hanging in clusters that are seen and mistaken as trees. In "well-fruited" yards and on the outskirts of the road are trees bearing mango, mamoncillo (guinep), ackee, and papaya. For the majority who travel to Negril, modes of transport can include personal cars, rental cars, vans, mini-buses, commercial buses, air-conditioned vans, and tour buses with television screens that all follow Route A-1 for the trip. All of this is on the way to Negril.

Highway A-1 meanders along Sandy Bay, through the town of Lucea and the village of Green Island then onto Orange Bay, where it was renamed "Norman Manley Boulevard." One of the early signs of Negril's tourism development was the naming of this final section of the road after the Rte. Honorable Norman Washington Manley, a national hero of Jamaica, one of the founding fathers of the Peoples National Party (PNP) and the modern Jamaican state. Passing Rutledge Point on the left is the Negril Aerodrome constructed in 1976. This small airport made the town more accessible than ever before for business travelers such as hoteliers, workers, and vacation travelers to this expanding tourist area. When Route A-1 passes Orange Bay, the end is in sight, as this is the most northern point of the renowned white sandy beach of Negril.

On each side of Norman Manley Boulevard are resorts and accommodations of all sorts and sizes as well as places to eat, drink, and shop. At Negril Centre, following the circular pattern of the roundabout, the road becomes West End Road, which leads to the cliffs and the lighthouse toward the west. Bearing to the left, the road becomes Nonpariel Road, which goes southeast toward other towns that lead to the parish capital of Savannah-la-Mar.

Regardless of ownership, all-inclusive resorts demonstrate "enclave" tourism development that is not "sustainable" with limited integration between "all-inclusive" properties and other businesses and the local community (Poon, 1988). This brand of gated community tourism prevents the tourist dollars from filtering into the wider community economy and impedes guests from getting the cultural experience at the destination. Outside of resort staff, they seldom interact with the community, which results in comments like that one from a female visitor found on TripAdvisor inquiring if it was possible for her to venture out of the resort property and if it was safe to do so.

Travel blogger Dan James in "Layer Culture" says that staying safe while on vacation should be everyone's priority and doing so in Jamaica is no exception. These words are meant to assuage the fears of travelers in face of the ongoing travel advisory issued by the U.S. Department of State, warning tourists to avoid secluded places or situations, even in resorts. Further, the U.S. Department of State cautioned its citizens, "Sexual assaults occur frequently, even at all-inclusive resorts" (Travel state.gov/content Jamaica

Advisory Level 3). In stark contrast, complaints have been trending down to the point that Jamaica Tourist Board data show that over the period 2012–2017, an average of only eighty-six complaints were filed annually by visitors. On top of the list of complaints over that time period were complaints of poor customer service, harassment, misrepresentation to visitors, and a lack of security in the accommodation sector, particularly small and noncompliant properties ("TPDCo urges compliance with Safety Protocols," Johnson 2020). The Tourism Product Development Company developed a presentation "Treat Our Visitors Right"—TOVR (TPDCo TOVR campaign, Renee Johnson; January 19, 2019). Reminders included, "Please, Thank you and Excuse me go a long way"; "Be authentic—no need for phony accents"; and "Respect Visitors' Personal Space—do not touch!!" Swedish ethnographer Orvar Lofgren (2004, 46) noted that there is a basically worldwide phenomenon of harassment on the beach where tourists and local people interact. Tourists complain about being hassled, or forceful selling by women and men who vend their wares on the beach. This has led to the policing of the shoreline by members of the local constabulary or resort security.

LINKS AND WEAKNESSES

There are a few issues that require attention in reference to Negril, its development, and the community at large. With the acceleration of investment in the town, there is an ongoing struggle of social and economic inequality that is exacerbated by tourism. It seems like "donkey's years" ago (actually early 1950s) that someone ridiculed a family member when he cleared a space on his beachfront land and erected changing rooms for visitors to use for a modest fee. The villager said, "Boy, wha 'yu father say him a do? Bring di touris' dem a Negril?" "Earl" replied, "Yes, sah." The response was "Ha! Ha! Him goin' fi hunger first, before touris' dem lef MoBay come a Negril" (ISER 2002, 71). Although "Earl" did not earn as much as he anticipated in the changing room business, he was able to sell parcels of the land to investors, providing him a good flow cash for family needs.

The population of Negril fluctuates, but as of 2019 there were 12,027 people employed directly in tourism (UNTWO 2019). The development of tourism in Negril pulled from a wide range of rural communities and there is a heavy flow of commuters coming from Little London, Grange Hill, Sheffield, and Green Island to name but a few. Finding adequate housing is a major issue for all, including the professionals who work in the industry. With few job opportunities available in the nearby sugar industry, unemployment is high. Although COVID-19 marks extreme measures due to the pandemic, the data is revealing. October 2020 data show overall unemployment at 44.00 percent for women and 40.2 percent for men. Pre-COVID-19 percentages show youth unemployment at 16.30 percent in 2019, which increased to 28.00 percent January 2021. *The Gleaner*, January 20, 2021, reported that the largest decline in employment during the pandemic occurred in the "Arts, Entertainment, Recreation, and Other Services, Accommodations and Food Service" sectors.

In Negril, the cost of living for those who rely on this seasonal work and fund those "slow" months when they would be without employment is highly challenging and documented by research (see Dunn and Dunn 2002b; ISER, 2001) and in *Women and Tourist Work*. For example, the minimum weekly wage in 2019 ranged from JM$5,500 to 6,200. Consider that wage factored in just food. The following numbers were gleaned from Numbeo (n.d.). Ingredients for an omelet: JM$145.00 bread, JM$180.00 eggs, JM$122.00 cheese, total JM$447.00 for one meal, not including the cost of butter and the utilities used to cook the meal. On top of that, Negril has a notorious 100 percent higher cost of living than the national average. Seasonal employment in good times as well as perpetual high unemployment that leads to harassment of visitors, petty crime, and other negative behaviors are factors in tourist-dependent Negril. In addition to these issues, there are violators of the very base of the environment, the beach and the reef.

CORAL REEF RESTORATION

There is a major threat of shoreline degradation. By the mid-1980s, divers began to notice the destruction of the reef. Anthropologist Barbara Olsen (1997, 287) recalled how dive operators and instructors responded to calls to form the Negril Coral Reef Preservation Society (NCRPS) to protect marine ecology. A few years later an awareness campaign took off that focused on diver damage and was distributed to all hotels, resorts, restaurants, and business establishments. Guest divers were implored to "look but do not touch" the reef. There were cautionary posters that explained the dire consequences that occurred in the sea and on the beach because the reef was dying. All who worked in that environment, such as fishing, snorkeling and SCUBA diving, and charter boat personnel championed the cause. The NCRPS received grant money from international lending agencies and fostered alliances with Jamaican and Canadian ecology and environmental groups. Environmentalist Janice Francis-Lindsay (2019, 156) suggested that balancing the needs of the tourism industry must be in tandem with supporters of the environment and the local community.

An example of community action was featured in a newspaper article (March 19, 2019). Twenty-five years after its founding, the Negril Area Environmental Protection Trust (NEPT) continues its efforts of rescue of, restoration of, and education about the reef. In 2019, the organization embarked on a major restoration project of Negril's coral reefs by establishing a coral nursery in the Orange Bay Special Fishery Conservation Area in Hanover. The nursery currently has approximately 1,200 pieces of different species of coral that were set up by the core team, which consists of six members, along with a crew of volunteers from the community and its environs in January. When the corals mature, they will "out-plant" off the coast of Orange Bay to restore damaged reefs to a healthy state and boost fish sanctuaries. The executive director of the NEPT, Keisha Spence, explained that over the years, the reefs in Orange Bay and Negril, by extension, have been experiencing continued levels of degradation due to external forces such as "unruly fishermen" and businesses situated along the coast. The NEPT, through support from the Environment Foundation of Jamaica, heavily promotes its community-awareness initiatives. Spence remarked, "It is one of the largest efforts in western Jamaica as it relates to coral restoration that uses an underwater drone for surveillance and documentary recordings of the coral nursery. Fishing is allowed only with permission" (*The Gleaner*, "Environmental Protection Trust Restores Negril's Coral Reefs" March 16, 2019, 1–4)

The primary challenge is to convert those who believe [their livelihoods are in jeopardy]. Fish sanctuaries do work, and they have seen the benefits, but you have some persons who feel they are displaced and so do fishing in these no-fishing zones. (ibid.)

In addition, the NEPT supported at least eight fishermen who also do farming as an alternative livelihood and protect the Orange Bay Special Fishery Conservation Area, which has a dense population of mangroves and seagrass, as well as small communities of staghorn and elkhorn corals that house other marine wildlife including sea cucumbers and stingrays. Significantly, the NEPT hosts local events to spread awareness and facilitates school visits of the area. Among its outreach duties was an environmental program for seventeen schools in western Jamaica, engaging more than 250 individuals. A local Orange Bay fisherman cited the benefits of the SFCA with the steady increase of the fish population. He called for more support of the NEPT in its effort to restore the coral reefs. "These people are trying to build back the reef, so they need support because we can't do it alone," he said (ibid.).

THE NEGRIL CHAMBER OF COMMERCE

Founded in 1983, the Negril Chamber of Commerce's stated mission is to facilitate the sustainable commercial, social, and environmental development of Negril. "It is the voice of the community in that its members strive to develop projects, protect and improve the tourism product and promote Negril as a tourist and investment location" (Negril Chamber of Commerce, Negril Guide 2020, 64). To this end, the work is voluntary and encourages its members to sensitize, educate, and lobby. Each year, save for the pandemic year of 2020, the NCC produces the *Negril Guide*, which illustrates what this group of hoteliers, tourist sector workers, and the community do. Moreover, the *Guide* illustrates how effective the NCC is in keeping Negril close to its ethos of a small, close-knit community who works together. Further, the NCC bridges tourism with tourists for their mutual benefit. The NCC received funding from government agencies for local programs under the UDC umbrella and lobbied the government for guidance and technical and financial assistance for infrastructure improvements. Each year in the *Guide* is a list of progress made toward the NCC key objectives for the year. In 2018, the NCC secured funding to bring two coastal engineers from the Netherlands to present environmentally sound solutions for beach erosion, distributed garbage bins around Negril in partnership with a NCC business member, sponsored International Coastal Clean Up day, and reopened the Negril Recycling Center including training members of the hotel industry about best practices of waste management. Among the list of ongoing efforts are lobbying for advanced discussions for waste management, improved water supply, safer road infrastructure, and implementation of a mini sports complex in an effort to lower crime and violence in Negril. In lieu of charitable fund raisers, such as galas for the Negril Health Clinic, the NCC now has a nongovernmental organization (NGO) arm called the Negril Trust Fund that will allow them to apply for grants to fund their projects pertaining to the social and economic conditions of the community. The Issa Foundation is poised to equip the clinic in service to Negril.

NCC is committed to the small businesses that were the start-ups in Negril in the 1970s. Often, the Jamaica Tourist Board promotional efforts did not apply to the small size of the establishments in Negril. Therefore, the NCC encouraged and helped small hotels and guest houses to produce their own promotion of their businesses on social media and are successful at doing so. "The Real Jamaica Experience" is another partnership with members where visitors' money goes directly into the hands of local Jamaican people who are providing goods and services. Hostels and individual rooms are available within local communities that are suitable options for "the Cheap and Cheerful" crowd. This service can be found online through Airbnb, and booking and reviews are noted on TripAdvisor. These endeavors illustrate the bridge that NCC maintains, not disparaging tourism, but negotiating its borders to improve the economy and the lives of those working in the sector.

COVID-19

By December 2019 Jamaica's tourist industry was poised to exceed its expectations of the upcoming season. After all, Jamaica saw a 3.1 percent increase the previous year to 6.5 percent. By March 2020, the GOJ closed its borders in the face of the spreading of the pandemic of the novel coronavirus, COVID-19. This action brought the economy to a standstill. Without tourism, businesses stood to lose the equivalent of US$3 million a day, revenue that sustains 20 percent of the country's economy (Meade 2020). By June 2020, Jamaica reopened business for tourism when the minister of tourism, Edmund Bartlett, announced a plan to establish a "resilience" or "tourism" corridor that runs from west to east along Jamaica's northern coast, where all travelers would remain. Vacationers must remain in the tourism corridor and not be quarantined, or if outside of the tourist area, they must enter fourteen days of confinement and have their movements closely monitored. Jamaica instituted a "travel authorization form" for entry into the country. Travelers from specific states, such as Florida and Texas and other hot spots in United States, undergo further scrutiny as well as those coming from Britain and other parts of Europe. Travelers must stay in the corridor. By the end of September 30, 2020, there were 6,408 confirmed cases, 1,624 recovered, and 101 deaths due to COVID-19. Jamaicans wear masks and understand the risks of this public health crisis after undergoing the 2016 chikungunya and dengue fever epidemic that crippled the country at that time. Mosquitos are one thing; COVID-19 is another.

In Negril, the pandemic hit hard because of the number of international visitors who brought the virus, and the lack of tourists whose dollars and Euros fuel the economy of the community. Not only was COVID disruptive of business, but also the resumption of business to normal levels was unknown. It was clear that because of business/tourism shutdowns, the livelihoods of the workers also came to a standstill. One early marker of the pandemic occurred at a boutique hotel on the Cliffs, which saw a dramatic number of cancellations, mostly coming from its European clientele. Further, sixty-four out of its seventy-six employees were sent home, presumably for the duration of the hotel closure (Durrant Pate, March 2020, Jamaica Observer p. 1–3.). The article stated that several other hotels in Negril were shuttered due to a lack of business, while the few that remained open significantly reduced their workforce. RIU Negril, which employed over 300 workers, announced that in early March the hotel would be closed. The gates were boarded by zinc fencing that was unsightly and unnerving during the height of the pandemic. There was optimism voiced by the leaders in Montego Bay's tourism that the industry would "ride out the storm." However, those who relied on work in the sector were set adrift without any economic safety net.

Through aggressive public health actions, the Ministry of Health, Immigration and Customs, and other agencies took control as best as possible to reduce the spread of the virus in the general population. In the meantime, Negril, like the rest of the "tourist corridor," saw tourist workers who were released from their workplaces or because of the lack of customers had little money for food for themselves and their families. Hotelier and philanthropist Paul Salmon used his talents and his long-term relationships with the people of Negril to establish a campaign to feed the needy of this tourist-dependent community. Using a button on the Rockhouse Foundation webpage, Salmon and his team solicited funds to feed desperate families. Targeting the families already in partnership in their education program network, they made weekly distributions of food staples to families in and around Negril. A safety net program aimed at delivering benefits to the most needy and vulnerable in the community steadily expanded to include neighboring village families who were also struggling to meet basic nutritional needs. The Rockhouse Foundation was making weekly food distributions and sustaining approximately 1,000 people in the community with food staples.

According to Sophie Grizzle-Roumel, an interlocutor in this study and the director of Charelea Inn, the hotel staff has an 85 percent vaccination rate—the highest in all of Jamaica. She commends the government for excellent COVID-19 protocol training and success in the Resilient Corridors that stem the spread of the virus across the island. Charelea Inn's website lists antivirus sanitation and other practices put in place since March 2020. For small hoteliers, the pandemic has been a life-resource-straining event and for some perhaps end of the business. Long-term returning guests who established personal relationships with Charelea staff sent money to them to stem the tide of underemployment during the year. Not all staff were fortunate. In an interview, Donovan White, director of tourism, remarked on the impact that the pandemic has had on the country—the performance of the sector was down some 68 percent in comparison to the previous year.

> In total, we had somewhere in the order of 850,000 arrivals for the year, compared to 2.5 million the year before in 2019, so it's a huge fall-off. The comparative fall-off in earnings would be somewhere around 73%, because our earnings for the destination were around $1.3 billion in 2020 compared to $3.8 billion the year before. (Nelson 2021)

Further he stressed, "The impact on the overall economy has been very stark" (Nelson 2021). Both Grizzle-Roumel and White expressed concern about the potential impact of virus variants plaguing Jamaica just as business seemed to see a glimmer of hope. White remarked,

We have to continue the process of vaccinating our population. Jamaica is one of the most tourist-dependent nations. . . . So, recovery for us is essential. The tourism industry employs some 170,000 workers in the industry. So far, we've only been able to return somewhere in the order of about 50% of that number to active work.

The workforce is of paramount concern for Mrs. Grizzle-Roumel not only as a hotelier but also as treasurer of the Negril Chamber of Commerce. She bluntly told her staff that it was an either/or situation: getting the vaccine or not eating, meaning that if they did not get the shot, they would not have the means to survive. Evidently, Sophie Grizzle-Roumel's words were heeded by others. Thanks to the diligence of members of the Negril Chamber of Commerce and their employees, Negril was close to herd immunity.

Chapter 3

Women, Work, and Tourism

This chapter looks at the complex nature of women, work, and family as played out in Jamaica and in Negril. To do this, the analysis required a critical "travel ready" intersectional lens that allowed the social vectors of class, race, and gender to guide the overlapping social connections found in the society. In the Jamaican context, those vectors pertaining to the role of women were established during the days of enslavement of thousands of African and African-descendent peoples. Considered next are ways the gendered system encoded British colonial concepts of femininity during the post-emancipation period. Moving forward in time showed how the majority of Jamaican women and men accommodated Eurocentric norms, constructing a model of their own norms and cultural values. As such, women were the economic providers for most of the households in the society. Finally, the chapter concentrates on women tourist works in Negril as micro- and small-business owners.

AFRICAN-DESCENDENT WOMEN
AND ENSLAVEMENT

As enslaved women, Jamaican women performed domestic labor for their own households and for their masters and in the sugarcane fields. During British colonial enslavement (1655–1834), as the eminent Jamaican feminist historian Lucille Mathurin Mair (1975) noted, the majority of the enslaved women and men were equal under the whip, cut sugarcane, gleaned the cane trash after harvest, and served as traditional medical practitioners. During this time, as Mathurin Mair (1974) was first to argue and document, women used "weapons" that were available to them to challenge the slave system. There were a wide variety of acts of defiance including insolence that occurred on a daily basis. As Bush (1990, 61) explained, "Unlike outright revolt, these unspectacular routine acts were resistance" to enslavement. Further, of all the enslaved, domestics probably exhibited the greatest degree of duality of behavior. Outwardly, they conformed to and adopted White culture to a greater degree than the more autonomous field slaves, while covertly rejecting the system of "grinning and bearing" with a personal edge to it. It is not just going with the flow but the acting out of disdain and agency in situations where a person does not have control over personal outcomes. The Guyanese poet Grace Nichols's poem "Skin Teeth" (1992, 797) expressed this sentiment most eloquently when she wrote that by "showing skin teeth on bended knee know it is the best to rise up and strike." This additional arsenal of traditional cultural practices helped women to mediate the inequities in face-to-face encounters. At the same time, a sense of Jamaicaness was a cultural production evolving from enslavement. Jamaicaness is a notion of shared respect and care extended to one another's benefit. One can show skin-teeth as a smile as a sign of endearment and in a joking manner among equals. In this fashion, Jamaicaness becomes so gratifying as it made the sociocultural conditions among the enslaved humane. On the other hand, showing skin-teeth with appropriate body language displayed displeasure, especially in an inequitable situation, as noted in the lines of Nichols's poem.

Another product of slavery was that women became mothers and marital partners by choice or force. Moreover, the progeny of White overlords and their enslaved Black women created a middle status "colored" group in the White/Black/Brown stratification system on Jamaican estates. Over time and under varying circumstances, light skin color and free or non-free status became important even during enslavement. According to Caribbean feminist scholar Cecilia Green (1999, 7), Afro-Caribbean women's manipulation of the raw material of their lives depended on three concrete things, among others: one, the presence of the model of Eurocentric-dominant gender role expectations was missing due to the high rate of absentee owners and their wives; two, the existence of autonomous spaces of social life; and three, the availability and invocable of an African cultural memory. These factors afforded the independent economic agency of Black women, evidenced especially by their access to land, their huckstering or higglering roles, high levels of household headship, and consistently high labor force participation rates. Further, Green (1999, 13) suggested that these opportunities for Afro-Caribbean peoples were solidified by the post-emancipation period, enabling them to form Black peasantries to retain, to reinterpret, and to reconstitute certain Afrocentric cultural principles.

Beginning in the post-emancipation period, women across class and color lines engaged in a range of livelihoods. The majority of women were still residing in the rural areas; some worked on sugar plantations, experiencing gender disparities in the wages they earned. Also, during this period men migrated to join work gangs to build the Panama Canal and to cut cane in Cuba, which netted relatively good pay that was remitted to domestic partners, wives, mothers, and family. Women were raising children; some were buying land (thanks to Panama silver remittances), building homes, and establishing villages. Some women were in business for themselves as higglers, seam- stresses, small agriculturalists, and laundresses, which afforded the funds to send children to school. If lucky in their circumstances, some women were able to become literate and engage in community and church affairs.

The archival study of Jamaican villages by Sidney Mintz (1974, 157–79) documented the free village movement founded under the church leadership of Baptist missionaries. "In this context, the minister became a substitute an altogether preferable substitute—of the estate owner, the overseer the slave driver, the judge, and the custos" (179). An underlining issue, however, cen- tered on gendered relations. The intention of the development of Black peas- ant villages was to create "a rural middle class of yeomen and cottagers with a stake in property, wage labor and Afro-Saxon respectability buttressed by the church, proper marriage and the spirit of thrift and self-help" (Green 1999, 17). As historian Catherine Hall (1995, 53–54) pointed out, "The missionar- ies envisioned no less than the transplantation of the solid, sober English middle class model into the rural Jamaican countryside." Subsequently, a new gender order was central to the vision of these abolitionists. In their new Jamaica, "Black men would survey their families with pride, black women would no longer sexually subjugated to their masters but properly dependent on their husbands" (Hall, 54). Not all were part of this stylized formation of rural Afro-Saxon communities.

At this same time there was an increase in size of the rural female labor force as well as a spike in the numbers of women migrating to the capital, Kingston. There, they sought employment as domestic servants in the homes of the growing urban middle class. If skilled, educated, and of light skin color, they found jobs in colonial offices or set up shop as dressmakers and as seam- stresses, to name but a few occupations available to them (Shepard 1999, 89). The goal of these women was to be as self-sufficient as possible and to support those who relied upon them by sending money to the rural homestead.

ENCODED VICTORIANISM

Thus, as Jamaica entered the twentieth century all the way through to the contemporary moment, Jamaican women strived to secure economic resources for themselves and for their families. Further, despite the domi- nant Victorian ideology that presumed that a woman's ideal position was to be legally married, the majority of women in Jamaica were mothers and economic providers with or without male partners. During enslavement, a licensed marriage was an impossibility. Consequently, households of the enslaved were organized as visiting unions whereby a woman was "visited" by her domestic partner, or as suggested by historian Barry Higman (1979), a significant number were in consensual unions, arrangements that were not a legal marital unit but for all intents and purposes were structured as such. Regardless of the legitimacy of a marital union, the births of children were welcomed. Following emancipation, these three marital practices took on their own cultural stamp of approval. Those most welcomed children were not given legal status if their parents were not legally wed. Besides religious affirmations, to enter legal marriage for many working class and poor required more than an affectionate marital commitment, but significant demonstrations of financial support and the funding of "a proper" wedding. Moreover, as UWI feminist scholar Christine Barrow (1996, 37) noted, for women, "marriage has the sanction of respectability and is also the hallmark of status," in contrast to how their men felt. For men, circumstances of poverty and unemployment might make them reluctant to take on those financial burdens of the "proper wedding" and investing in a home. Often, it took a decade or more for a potential groom to fulfill these obligations. In the meantime, life goes on and children are born.

These domestic arrangements have social and cultural underpinnings. For instance, there was "The Mass Marriage Movement" (1944–1955), a colonial government campaign organized to convince couples to marry. The goal was to encourage couples to legally marry, thereby legitimizing their offspring. This campaign came on the heels of returning World War II vets and migrants returning home to Jamaica. Jamaican feminist cultural artist Honor Ford-Smith (1988) called this the "housewifisation of women," resulting in the displacement of women from the workforce. On the whole, the Mass Marriage Movement did not deter the masses from continuing their cultural practice of common law marital union, or women giving birth to children without sanctity legal marriage. Children were welcomed but were still considered illegitimate in the eyes of the courts of law. This issue was ultimately put to rest with the passing of the Status of Children Act of 1976. Finally, Jamaican children were freed from the legal stigma of illegitimacy regardless of the marital status of their parents.

With that said, no matter where they are located in urban, rural, town, or village settings, there is a range of domestic arrangements in Jamaica whereby sometimes men visit their children and domestic partners at different points in their relationship (Barrow 1996) and many of those households are headed by women. Socially and culturally fluid and usually limited by small physical spaces, household membership can include fathers, women kin, male kin, cousins, and others (Bolles 1996). The key element that remains is that a woman who heads a household is both caretaker and economic provider. Confirming this position, Gina Ulysses (2007, 44) commented, "Working class women hold children as their primary responsibility and focus" no matter where they find that work.

WORK AND TOURISM

In the chapters of this book, I contend that tourism in Jamaica is a predominately "female" industry in terms of a product resulting from the expansion of neoliberalism that garnered capital private and public investments to the sector. The workforce in tourism is composed primarily of working-class women who rely on the sector for their livelihoods to finance the welfare of their children. Consequently, the precariousness of this business, hindered by hurricanes, global economic downturns, as well as shifts in political largess, hovers over the monetary opportunities for the women in Negril. Further, often the overdue material gains for women are overshadowed by gender systems that continue to be inequitable in terms of the status of women. As previously mentioned, the tourist sector is the second-largest employer of women outside of categories of "own account" or self-employed. Furthermore, many of those self-employed women can be found working in the tourist sector as roadside venders ("higglers," traditional agricultural producers and sellers) and sex workers, who usually do not get counted in official documentation. All of the social hierarchies and social differences of the society surface in tourism. The Jamaica Tourist Board amends its promotions to deflect Jamaica's major problems such as crime and drug-related violence by promotional tags, such as the early 1980s of "Come back to way things used to be," which harked back to a legacy of "colonial bliss" that is romanticized for tourist consumption.

As a service sector, tourist work fits neatly into Jamaica's economic hierarchy. Tourism's gendered segmented labor market provides service work in establishments that pay taxes, such as hotels, banks, and restaurants, that are counted and considered a service-producing tradable in the parlance of the GDP. Work categorized in the "elementary occupations" rubric is often not counted, but is critical to the tourist business, such as roadside vendors and sex workers. And then, there is the overall image women's tourist work as the chambermaid and the sex worker that solidifies the situation. In her classic text *Bananas Beaches and Bases*, Cynthia Enloe (1990, 34) remarked how in the 1970s, Caribbean nationalists complained that their government's pro-tourism policies had turned their society into a "nation of busboys." Sarcastically, Enloe suggested that perhaps "'the nation of chambermaids' did not have the same mobilizing ring in their ears." After all, a woman who has traded work as an unpaid agricultural worker for a hotel cleaner has not lost any of her femininity. This quote brought to mind two points. First was the concept of femininity that was aligned with the standard Eurocentric notion of what was considered a gender norm. Second was the assumption that the labor of a working-class woman was deemed of such low value that no matter the location, it was a gender-normative occupation. Nevertheless, the success of the tourist industry is categorically based on low-waged female workers. With the explosion of tourism in the Caribbean and particularly in Jamaica, the work of this "chambermaid" became the bedrock of the industry. Furthermore, if success in tourism is engineered by good service, hospitality, and professionalism, then the role that women play was critical to that accomplishment.

RACE, CLASS, AND EDUCATION

Before going further, there are things to be said about the Jamaican social system that are based on the interconnections of skin color, class, and access to education. In the Jamaican context, skin color (phenotype) and class are historically connected with people of darker hue (the majority) occupying the low strata of society. However, a proven key toward upward mobility in Jamaica continues to be education. Accordingly, following independence in 1962, the number of citizens with an elementary education increased dramatically when primary school up to grade 6 was made compulsory. Although technically free, secondary education remained costly for low-income households. It was not just the "school fee" but the required uniforms, school supplies, registration, and examination expenses that drove up the cost. In 2015, the bill for one child in secondary school amounted to US$300. That figure accounted for almost 20 percent of a Jamaica per capita income. Those are just some of the contributing factors responsible for the high rates of students dropping out of secondary education (Trines 2019). Although there was a high (97 percent) participation rate in elementary schools, 2017 data showed that the percentage dropped to 60 percent enrolled in secondary school education (Trines 2019). Those who advanced their education through technical training and university levels illustrated how wide the gap can be. Plus, access to education continued to be heavily skewed toward higher-income households in the Kingston Metropolitan Area (ibid.). In the 2010s, The Government of Jamaica (GOJ) contribution to education rose from 3 to 6 percent of the GDP and by 2017/2018 increased an additional 3.9 percent. Caribbean Advanced Proficiency Exams (replacement of British "A" levels) and Technical and Vocational Training (T-VET) credits provided entry into skilled, and managerial rank employment. An associate of arts in tourism earned at a T-VET institution, such as the community college in Montego Bay, became a valued credential. There are five levels of achievement in the T-VET program with the managerial and professional work at the apex. In 2017, there were 66,000 students enrolled at T-VET institutions (ibid.).

Other obstacles still faced by women include skin color gradation, which harks back to the days of slavery but with a contemporary twist. Women considered "Brown" can be light-skinned with somatic features (light eye color, hair texture) or dark-skinned. *Brown* as a social term also referred to middle-class status. In 1938, Zora Neale Hurston (1981, 76) made this observation, "If a woman is wealthy, of good family and mulatto, she can overcome some of her drawbacks. But if she is of no particular family, poor and black she is in a bad way indeed in that man's world." In contemporary times, the "color bar" can be lifted as new economic and educational opportunities become more available to a larger number of women than before. However, the options are fairly limited for the majority of women. Since these women are poor or working class and Black, the class/race nexus is further apparent.

The combination of class and color constraints is historically based but held tight in Jamaica's highly stratified social system. A seminal work on social mobility in Jamaica by Derek Gordon (1989) was prophetic when he noted the following:

> researchers must confront the paradox of large scale social mobility generated by the opening up of new positions coexisting side by side with gross and perhaps, even widening inequalities of opportunity between the minority at the top and the majority at the bottom of the social order. (Gordon 1989, 47)

In sum, class and color do count in factoring mobility; however, these social factors also intersect with gender. As a host of Caribbean scholars and their feminist colleagues from abroad argue, sexism constrains and limits women's access to opportunities in Jamaica. The ongoing social and economic struggles faced by women in Jamaica also have a history of those who persevered despite these obstacles. An example is none other than the former prime minister of Jamaica, the Honorable Portia Simpson-Miller (2006–2007; 2012–2016), and other outstanding women who are noteworthy in a variety of fields of endeavor throughout the society.

TOURISM AND WOMEN WORKERS

Early on in the tourism research on women, studies focused on the hosts and new workers in development programs. Margaret Swain (1995) brought the importance of including women housed in a gendered frame as members of communities that are now tourist sites. For example, Mary Castelberg-Koulma (1991, 197) noted, "Few studies address themselves to the effects that tourism has on the local female population." Or as Dallen Timothy (2001, 246) reminded, "Gender differences are obvious in the production of tourism in developing countries . . . and women's work resembles the types of work they have traditionally done in the domestic setting." Heather Gibson (2001, 28) reviewed a body of literature, for example, Kinnard, Kothari, and Hall (1994), which focused on "gendered hosts," females confined to jobs such as chambermaids, laundresses, and waitresses, usually characterized by low pay, low-job security, and powerlessness experienced in those tourist sites. Kinnard and Hall (1996, 96) argued that any gender analysis should address the differences in the quality and the type available; the differential access of women to employment opportunities; and seasonal fluctuations and the existing anew gendered divisions of labor. Ten years later, the focus on women as workers and as members of communities that are now tourist sites was exemplified by scholars like Kristen Ghodsee (2005) writing about Bulgaria or Denise Brennen's (2004) work in the Dominican Republic, or Nandine Fernandez's (1999) study on race and tourism in Cuba. Those bodies of work not only examined the factors pointed out by Kinnard and Hall but also went further by putting those traditional female jobs in context of a complex of opportunities afforded to women, and how women instrumentalized those activities for their own advantage and understood the consequences of those actions.

Change comes slowly even when material conditions make it possible for women to have greater access to resources in society than ever before. More importantly, regardless of their class position, when a woman's wage is combined with that of her male spouse/partner (who, more than likely will earn more than her), it creates a dual-income household that often generates greater earnings for that domestic unit. In Jamaica, Dunn and Dunn (2002a) found there was a distinctive disparity in income earned by men and women, showing that males were three times more likely to have a higher income bracket than females. As of 2020, that wage gap increased so that in Jamaica, for every US$100 a male earned, a woman earned US$ 60.00 (O'Neil 2021). More often than not, there were very poor women who eked out a very minimal existence. The overall condition of living in abject poverty was perhaps made more acute in the face of the tourists who ignored stretched-out hands as they begged from the side of the road or in front of supermarkets.

The UNWTO joined forces with UN Women, other UN agencies, and organizations to compile data and information concerning the contribution that tourism was making toward the UN Sustainable Development Goal 5—Achieving Gender Equality and Empowering All Women and Girls. The second edition of this report looked at key factors that have a say in gender equality in the tourism sector and that pinpoint challenges and identify ways to lessen inequality while capturing the potential that tourism offered to women participants. There were five thematic areas—employment, entrepreneurship, education, training, and leadership—all demonstrating what gender equity and women's empowerment in tourism would look like (UNTWO 2019, 1). Case studies provided four key objectives for tourism industries—digital platforms and technology, hotels and accommodations, tour operators, and community-based tourism. UNTWO (2019) reported that the majority of the global workforce in tourism is female, whereby 54 percent women were in the sector while 39 percent were in the broader economy. Further, the wage gap was indicative of the low-skill work usually performed by women, thus earning 14.7 percent less than men. However, tourism offered women great opportunities for leadership roles, as 23 percent of tourism ministers were female compared to 20.7 of all government ministers overall. The UNWTO report noted three major points for discussion: (1) women were challenging gender stereotypes in all sectors and entering occupations usually dominated by men; (2) technology was an important factor empowering women when coupled with training opportunities that encouraged women's entrepreneurial entry into tourist markets; and (3) policy-makers were becoming more aware of gender equity in tourism by putting measures that helped to ensure women shared in the benefits that tourism can provide.

The UNWTO report is an important referent for nations just coming into the global community of tourism. However, Jamaica, considered a mature tourist destination, has at least a 125-year history experience of being in the business. Nonetheless, there are still critical sticking points that remain echoed in the UNWTO report. These issues include the continuation of opening up of employment across all subsectors of tourism, using technology and the digital environment as an avenue for women's empowerment as entrepreneurs, and the long-term emphasis on gender equity not only in tourism but also throughout society. At this juncture, all of these issues are clouded by the 2020/2021 COVID-19 pandemic crisis. Secretary-General of the United Nations Antonio Guterres said, "In the midst of a global pandemic, one stark fact is clear: The COVID-19 crisis has a woman's face" (UNTWO "News Release" March 8, 2021). UNTWO Secretary-General Zurab Pololikashvili added,

> Tourism is a proven driver of equality and opportunity. This unprecedented crisis has hit our sector's women fast and hard, which is why gender equality and empowerment must be centre stage as we work together to restart tourist and accelerate recovery. (ibid.)

Policy-makers were encouraged to address women's economic and social insecurity in the face of the rise in unpaid care work and domestic violence (ibid.). The Government of Jamaica's Ministry of Labor (International Labor Organization) provided General Grants and Tourism Grants for small businesses, craft market vendors, beauty therapists, and others if they had applied for status through the Tourism Product Development Company (TPDC). How effective those grants were for women who own or work in those enterprises will be a topic for future study.

HIGGLERS AND ENTREPRENEURS

Like other Eurocentric models addressed by the UN, much of Jamaica's normative gender system is predicated on the dominant ideology whereby men are breadwinners and women are homemakers. One of the UN measurements for women's empowerment centered around women's entrepreneurship as an avenue toward gender equity. For Jamaica, entrepreneurship was encoded during the days of enslavement where Afro-Caribbean men, but mostly women, grew provision grounds and traded those foodstuffs in the internal market (Mintz 1974, 180–213). However, due to that same system, there was a crack in the male breadwinner myth in which today's Black women, the descendants of the enslaved, are seen as astute businesswomen. Writing about contemporary Barbados, Carla Freeman (2014, 1) wrote,

> *Entrepreneurialism* . . . is not simply a mechanism of self-employment—a vehicle for income generation, an economic matter of business, that is entrepreneurship in a narrow sense—but a subtler, generalized way of being and a way of feeling in the world.

As argued meticulously and artfully by Carla Freeman in *Entrepreneurial Selves,* the role of women's entrepreneurship and small business is very much part of Barbadian society coming out of the seventeenth-century British plantation system. Women's entrepreneurship morphed into an avenue for Black upward social mobility and is part of the ever-evolving neoliberal capitalist investment venture. In this study of Negril, tourism was an already established neoliberal development that framed the local economy. Therefore, the emphasis in *Women and Tourist Work in Jamaica* is not on the connection of middle-class respectability as an appendage of entrepreneurship or on global neoliberalism. What the discussion looks at is the history of higglers as micro-entrepreneurs and how this tradition is now used as a business model for poor and working-class women in Negril's tourism. Further, this long-standing tradition is utilized by Negril's women entrepreneurs because it relies on access to consumers, a distribution system of goods and services, strategic use of time, and inventory management. Also, Jamaican higglers are called just that, while in the Eastern Caribbean (see Carnegie 1986), they are referred to as hucksters and traders.

LONG-STANDING TRADITION

After the emancipation of the enslaved, the wider society considered higglers in a negative light. Collectively, higglers were fixed as being shrewish, loud, boorish, aggressive, and exhibiting behavior associated with lower classes of society. In the Jamaican context, this assumed skin color too. Skin color was a critical factor, as the majority of the working class and the poor are dark-skinned. Part of the higgler identity also had to do with what was deemed socially acceptable deportment for women. Regardless of the business acumen, keen awareness of trends, and financial soundness, higglers were not deemed proper. In recent years, feminist social scientists and activists have raised the status of the higgler in the eyes of many in the society. A case in point of new-age higglers is Gina Ulysses's *Downtown Ladies* (2007), an award-winning ethnography of these informal importers. Nonetheless, the contemporary small businesswomen, who are not higglers, are still tainted by the traditional image of market women, due to their aggressiveness. These modes of behavior are deemed smart and tough when attributed to a businessman. Businesswomen may engage in other kinds of commercial and business services that are highly technical and capital intensive but still remain in inequitable positions due to the general ideology regarding women in the economy (Carter and Cannon 1992). Of course, there are women who have been extremely successful in business, and Jamaica has quite a number of them too. Each of those women's rise to the top included surpassing not only basic obstacles but also those that were gender related, such as proving way beyond reasonable doubt that they were worthy, despite being a female. Again, this perception relates to the contradictory nature of inequality inherent in the gender system.

TAKING CARE OF BUSINESS

Evidence from both UNWTO reports and research from the University of the West Indies indicate the direction that governments, stakeholders, and women themselves need to embrace in order to successfully compete in today's tourist marketplace (UNWTO "International Tourism Highlights" p. 2. Institute of Social and Economic Research/OAS Research Project Impact of Tourism in Jamaica, 1991). These themes targeted the importance of technology, education and training, leadership, service, hospitality, and

professionalism to ensure women's empowerment. The narration of the interlocutors featured here in *Women and Tourist Work in Jamaica* exemplified
the extent that Negril is moving forward, and in some instances leading the
way regarding the UNTWO themes. Given that Negril was late in coming
into heavy neoliberal investment schemes, the time period of this twenty-
plus-year study is illustrative of tourism development as an ongoing process
in Negril. Whenever possible in the text, a specific point in time is indicated.

INTRODUCING THE INTERLOCUTORS

One of the goals of this study was to elicit interviews and carry out conversations with women tourist workers across wage-earning categories. All the
names and places are pseudonyms unless indicated. Listed are seven groupings that indicated the overall type of work and the number of women who
did that work:

1. Entrepreneurs (8): traditional higglers, dive shop, mini-store owner, hair
   braiders.
2. Cottage and small hotel owners (4): all owners of cottages and small
   hotels.
3. Professionally trained (4): accountants, clerks, hospitality desk.
4. Housekeeping (3): housekeepers.
5. Food and beverage (4): cooks, chefs, table captains, food and beverage
   managers.
6. Entertainment (6): activity directors, professional dancers, tour desk.
7. Craft vendors (2): general and specialty craft vendors.

Indicative of the gamut of occupations, the educational level ranged from
grades 3 to 6 in elementary to associate degrees to bachelor's degrees, a
BFA, and almost an MBA. The ages began at sixteen until seventy years. For
some workers, their jobs were learned as part of their socialization as girls in
their natal homes, such as the hair braiders and cooks. The second-generation
hotel owners worked in their family business early on in life. Good service
and hospitality—meaning more than just rendering service—were foremost
in the actions and deeds of the thirty women who participated in *Women and
Tourist Work in Jamaica.*

There were minimalist methods of inquiry used in the study. Potential participants in this research were contacted through verbal exchange after the
purchase of a good or service, with interpersonal decorum creating a sense of
trust between the two parties. Snowball methods of meeting someone on the
basis of another acquaintance were used as much as possible, again earning

trust and understanding between researcher and interlocutor. Sometimes it would take more than one encounter to set up a time for conversation and more often that did not happen. During those moments when an interview was not possible, observation became the key component, as just listening to the banter with others and watching interactions with consumers/visitors was a very valuable exercise. Being a fly on the wall was a solid ethnographic practice. Whenever possible, appointments were made and time allotments for interviews were kept to a minimum because these events took place in between paying customers or after work when it was a slow day or slow night depending on the job. This went on for years, as attempts in catching up during short visits, receiving details from friends conveyed through personal conversations, and during the pandemic on WhatsApp.

Another goal of this study was applying an intersectional lens to examine how women's choices or lack thereof were hindered by social obstacles such as lack of education or skills, and then what they thought of tourism as a way of making a living. Two things came to mind when drawing up a list of work categories. First was the fluidity of a job and then confluence of jobs. For example, among the four small hotel/cottage owners, all at one point in time served a housekeepers, accountants, and more for these family businesses until they took over the establishments due to their parents' or their grandparents' retirements. All of the women workers benefited from OTJ training, with some skills requiring more diligence in the business on top of formal training than others. Finally, the professionals among the group earned their postsecondary education in Jamaica, the United States, and Britain.

Depending on their age, many of these women tourist workers were leaders in their fields of endeavor. One was one of the leading housekeepers in her large resort and was in charge of the duties of her coworkers and was paid extra for this position. The table captain by virtue of her job description was the overseer of evening dining waitstaff. Further, among the interlocutors were community leaders who were well-established activists, particularly in the governance of Negril's tourist industry by way of the Negril Chamber of Commerce. In addition to NCC attention to business affairs, a cadre of women are members of teams that are concerned with recycling, cleaning up the roads, and marine life management (details of those efforts were explained in chapter two). Of the thirty participants in this study, two were native to Negril, but the majority commuted from nearby villages of Green Island, Lucea, and Sheffield. All actions are gendered as well as classed, raced, and intersected with other forms of difference. The notions of respect and reciprocity were central in this work; acts of Jamaicaness prevailed on both sides. It is evident that over the years when a return trip afforded a moment to catch up and to follow up with an interlocutor, the moment was welcomed.

MEANINGS OF WOMEN, WORK, AND TOURISM

The historical record shows the precedence of Jamaican women being in charge of the economics of producing, providing, or managing the resources essential to meeting the daily needs of their families. Children are welcomed and are the primary focus of a mother's work endeavors. In Jamaica, 45 percent of women are the primary providers (Global Information Society Watch, Jamaica, Dunn and Dunn 2013, 4–5). On top of that, for the majority of women there are social obstacles to overcome that are often grounded in the past such as skin colorism and lack of access to resources including education beyond elementary level, thus curtailing employment opportunities. Using an intersectional lens, it was clear that the impacts of race/color, class, and gender were intertwined. With 62 percent of the female work force (statin. jm) engaged in low-skilled tourist work, it is evident that this sector opened up jobs for poor and working-class women. After all, job creation was the basic premise in tourist development in Jamaica and elsewhere in the Caribbean. UNWTO directives outline the importance of job creation as an avenue for women's empowerment. However, as a mature destination—Jamaica's tourism coming on stream was full blown in the 1950s–1960s—the sector's solution to women's economic status and worth is an ongoing process, as evident in this study of Negril. In different kinds of tourist jobs, Jamaican women's self-worth and ways of surviving with dignity are truly tested. To maintain that sense of self, cultural practices are often invoked such as skin-teeth as well as asserting one's Jamaicaness so the guest is satisfied with their experience and returns.

Chapter 4

Welcome to Negril

Our Paradise on Earth.

—Negril Chamber of Commerce 2020

Women and Tourist Work in Jamaica turns to a significant feature of Negril, that is, the range of accommodations available to the visitor. The variety of places to stay harks back to the early days of tourism in the village. Now, those family-owned small lodgings share the beach with big-name mega-resorts. The range of places to stay in Negril meets a tourist's own interests, pocketbook, and sense of adventure. Naturally, in all of those properties there are women workers who serve and guide the vacationer on this journey. Since the emphasis in *Women and Tourist Work in Jamaica* is on the woman worker, her perspective on her face-to-face encounters with tourists is key. Personal service is critical to the success of the tourist industry, and certain cultural mechanisms are at play to ensure success by the worker and referred to as cultural practices of "Jamaicaness." To show "skin-teeth" is a performance among friends and family as a form of in-house joking. It is also used on behalf of the worker to assuage uneasy situations of inequitable social difference. Both practices are rooted in Jamaican culture which rose out of centuries of enslavement and almost 307 years of British colonial rule. Other issues arise concerning how women who own lodging establishments and/or are employed in hotels and resorts maintain their professionalism, regardless of the type of work they perform. In this chapter, the answers to those questions and the use of cultural practices are voiced by the women tourist workers themselves. They work in a range of accommodations, and their narratives are captured in notes, direct quotes, and in composites of voices, representing some of the women who make their living in tourism in Negril.

THE WEST END AND THE BEACH

One of the distinctive areas of Negril is the Cliffs or the West End. The Cliffs consist of meandering limestone that developed over 40 million years ago during the Eocene era (Robinson and Hendry 2012, 3). The outcome of this geological formation are these Cliffs, which consist of the occasional caves, inlets, and outcroppings with overhangs that face the sea. To reach the West End, the road begins at the round-about at Negril Centre going West for a stretch of 5 miles and hugging the coast. From its beginnings, the West End road was a rocky, white lime, dusty thoroughfare that is now a fairly smooth paved surfaced. Part of these improvements came over time with the further development of series of cottages, small hotels (some very high end), restaurants, and bars located on each side of the road. However, it is the history of the West End establishments that resonate with people. Some of the original lodgings are available for vacationers by Negrillians, including members of the Rastafarians communities who are also residents of the area. The West End implies laid-back relaxation and being close to nature with the backdrop of greenery on one side of the road that interfaces with the crashing waves of the sea on the other. In recent times, small hotels were erected on the precipice of the Cliffs. The living spaces overlook the outcropped spaces of sand littered with coral pieces with the endless background sound of the surf. Since the West End is on the western tip of the island, the sunsets are even more glorious as there are few lights coming off of the land.

It was here in the 1960s on the West End that urban middle-class Jamaican families traveled from Kingston for a relaxing vacation "in the country." Dotted along the West End road are cottages. Some of these cottages are part of a hotel compound, and often owned and operated by family members. Originally, families in Negril built extra rooms off of their own homes, or next door to them to accommodate Jamaicans or their own families on a holiday. This group tended to be repeat visitors. These lodgings were not luxurious, but very family-centered, and very welcoming, particularly on the part of the woman of the house, who was in charge of the visitors. Soon families were building cottages with Jamaicans and those from aboard in mind. Over time, amenities came too.

ON THE CLIFFS AT WEST END COTTAGES

A traditional family-owned set of cottages on the West End is illustrative of the kind of lodgings available on the Cliffs and the women and their families who own and manage these properties. Mrs. Cuthburt was educated locally but attended secondary school in Montego Bay. She left school early and did not sit the Cambridge examinations, the required entrée to good jobs and tertiary education. Her deep-colored eyes and wide smile highlight Mrs. Cuthburt's sienna-colored complexion. Her route to success was her marriage to a man whose steady job provided a cushion, and they were able to save. The savings were invested into their money-making cottage business. Mrs. Cuthburt took the time out of her schedule during a lull in the tourist season (around early June 1999) to be interviewed. These are notes taken from conversations that took place on Mrs. Cuthburt's veranda of her own house.

Mrs. Cuthburt owns and manages three cottages. She is the sole owner of one cottage, and the other two she owns with her husband. Mrs. "C" started out renting cottages to Jamaican middle-class families who vacationed in Negril in the early 1960s. She did all the cooking, laundry, maid service, desk clerk, "every little striking thing." Mrs. Cuthburt saved her money as business was good because of referrals and returning guests, some of them Jamaicans, but more from Canada and Europe. She acquired the two additional cottages with her husband's financial contribution provided by wages as a low-level manager on a nearby sugar estate. Mr. Cuthburt built the cottages one by one after work and on the weekends. Each has a veranda and a bedroom and a little alcove that accommodates a child's sleeping arrangements. Amenities now include a television and hot- and cold-water bathing facilities. Mrs. Cuthburt bore five children. All of her children played with the guests, and the guests' kids. They also helped their mother change linen, clean rooms, and run errands. Since they are now all grown up, they live in either the United States or Kingston with their own children.

Mrs. Cuthburt begins her workday as she has for almost forty years. She prepares an early morning cold breakfast spread for the guests. It consists of just juice, coffee, tea, toast cheese, and marmalade. After she sets out the trays on a table under the tree in front of the cottages, she checks the register to see if anyone is checking out or coming in. No alarming voices remind people to check out on time, but to be mindful when the time comes and to ask for assistance. Over the years, Mrs. Cuthburt has gone from doing it all— chief cook, maid, accountant, and the like—to hiring help. Now she employs one woman to help in cleaning the cottages and an old man to do the gardening and handyman work; she still takes care of the books herself.

As a member of the Chamber of Commerce, Mrs. Cuthburt worked hard to improve the infrastructure of the West End in regard to resurfacing the road. Waterlines are still listed as a future project. Her careful management has made the three cottages a successful business.

In 1988, when hurricane Gilbert, a Category 3 storm, ripped through the western part of the island, demolishing buildings and roads in Negril, Mrs.

Cuthburt was able to secure a rebuilding loan very quickly due to her relationship with the local bank. She makes daily business transactions with the bank and is known by the manager through Chamber of Commerce events. As a matter of fact, Mrs. Cuthburt has investments in other businesses around the island. As profitable as they are, none of her children are interested in enterprise of the cottages. However, she hopes one of the grandchildren will take over when Mr. "C" retires.

ON THE BEACH

Ethnologist Orvar Lofgren (2004, 38) said there are three basics that make up the global beach: sand, sun, and sex. That combination sums up the imaginaries of visitors who seek out Negril especially lured by the seven miles of white sand beach. Geographers list Negril's beach as one of the "world's best" and anyone who has seen it or walked on it knows that the claim of "best" is well-deserved. The goal of the tourist worker then is to assure that the vacationer has access to at least two of those goals, sand and sea—while self-indulgence is up to the couple or individual, advice about avoidance also is shared about sex if that is included in the inquiry. Because this is Negril, there is a range of accommodations available on the beach, across the street, and near Negril Centre. There are small hotels, family-owned establishments, and resorts. Since 2001, large international chain resorts have claimed beach space, so much so that a tongue-in-cheek comment cited by Stupart and Shipley (2013) remarked that this was "the second Spanish conquest of Jamaica." In that same year, RIU, a Spanish conglomerate, entered Jamaica's tourist sector and built its first resort in Negril. This chapter first considers small hotels that are family-owned and managed by daughters and granddaughters who were raised in the tourist business in Negril. Further on are conversations and observations of women interlocutors whose workspace is on or close to the seven miles of white sandy beach. Again, all of these conversations and observations came from my notes and all of the names of people and places are pseudonyms.

THE DAUGHTER

In July "Ms. Regina Hakkem" let me walk with her as she prepared for the day in her family-owned hotel, "Trees." Facing the other side of the Beach Road, Trees opens up with a luxurious driveway framed by Royal Palm and, now in bloom, Flamboyant trees. Interspersed between the trees are orange, yellow, and red Jamaican croton, and other flowering bushes. The property is organized in a semicircle with the buildings with rooms facing inward. Elegantly placed bougainvillea, hibiscus, and other floral gardening displays frame entrances to rooms, the restaurant, public spaces, and the office, outlining the many walkways of the property.

Ms. Regina Hakkem is a twenty-eight-year-old hotel manager who holds a bachelor of science degree in hotel management from the University of Miami. She was raised in the business; her parents also own a farm that produces most of the fruits, vegetables, eggs, and chickens for the hotel. Her brother became a physician, so, in order to keep the hotel in the family, Regina went abroad to study hotel management. She then came back home to take up her legacy. Single and not looking for a husband at that time, Regina is a dutiful daughter of Lebanese-Jamaican origins. This is what she had to say about her work, her life, and the future of tourism in Negril.

My day begins whenever a crisis interrupts my sleep. A broken water pipe at midnight means the day starts at midnight. One time last year, it was not a hurricane, but a storm, I worked two days straight. We had guests! It was a mess! The hotel is very open—it's natural Caribbean style—so when the wind blows and the rain comes, it can cause problems, especially for Americans. Europeans seem to handle things a bit better. So, take yesterday: my day started not bad, around 8:30 a.m. I live on property, but my apartment is not exactly on top of the guests. One of the perks, you might call it, is that my breakfast is prepared by the kitchen. I take it here right on this balcony [off the office]. One of the things I missed about home (when I was in Miami) was Blue Mountain coffee, but sometimes I do not get to finish my coffee before things happen (here). Yesterday my day was regular for summer season. I go over arrivals and departures of guests, meet with housekeeping and grounds staff. I do the food and beverage management, so that has to be dealt with, too. By then it's way past midday. Usually, I walk the grounds and look at things. I speak with guests, and if there are repeat visitors, I take extra time. Before tea, I work with accounts, make sure our bills are paid and that sort of thing. Of course, if a problem comes up, which usually happens, everything stops and I have to look to it. Like the other [day]—you noticed that we are building on—well, one of the pieces of machinery broke down, stopping all work. The contractor called a meeting and we had to reschedule the next week and the look at the timeline for completion date, which was behind time already. Then I had to recalculate the financial situation and so forth. This [is] low season, so at least there aren't a lot of guests to get in the way.

My day ends about 9 or 10 p.m. If it's a Monday, we host a welcome reception for guests, and that is a great social event—rum punch, cocktail patties, cod fritters, you know, a real Jamaican thing. I am on duty until it's over and then some.

Do I have a social life, being single and all? The answer to that question is no. One day, but not now. Maybe I will go on holiday for a week or so in September, do some shopping, visit family, but nothing else. Yes, it is hard to meet someone in Negril because I am always working. Where would I meet someone? Everyone is in town [Kingston], not here. The construction must finish.

I am fortunate that my father bought this years ago and built it up. I like doing this work. When I went to the States, I thought that I might like it and stay. And I thought the studies would be easy because I learned the hotel business as a child. What a shock! I missed Jamaica, even in Miami, and I realized too, that unlike my classmates, I did not have to look for work after graduation—one was waiting for me.

Tourism has really gotten stronger here in Jamaica and in Negril; things are growing fast. Soon there will be no idle beach land; you know the problems of the reefs. What worries me the most is the sanitation problem needs to be dealt with. We passed the code for the new construction, but I was in conflict the whole time. My father spent extra money and time to make sure that we were way above the standards, for our sake and because we [think] of the community, too. I spent my summers here and know the people.

Tourist development should not be done without thought. Those days are finished. I am in business to make money for sure but not to spoil everything, too. The government of Jamaica must work on that on a community level. But you know Jamaica—a lot of talk no action.

ANOTHER DAUGHTER

Some of the family-owned businesses are not native Negrillians. However, they have long-term connections with the community. Such was the story of "Marie," who spent her summers in Negril on her family's hotel property as a child, but now finds full-time employment there too. Marie was the food and beverage manager of a large hotel close to Negril Centre. She graduated from a prestigious girls' high school in Kingston and attended the University of London in the UK. When Marie returned home to Jamaica in the late 1970s, she became a Rasta. Rastafari is a religion and accompanying lifestyle that professes the importance of being African, being closer to nature, and revering Haile Selassie, late emperor of Ethiopia, as the messiah. Nowadays, people know about Rastafari by the music of Bob Marley, and there is a Rasta community in Negril, founded in the early 1960s. Nevertheless, after much prodding by her father, Marie moved to Negril to work in his hotel, located on the beach. She was hesitant to enter the tourist sector given her religiosity but found out she enjoyed the business. Marie worked hard as food and beverage manager. In that job, she had to procure foodstuffs for the tastes of her American clientele. Her main objective was to buy from local farmers, all at a good price. Through her networks, Marie was also able to support her Rasta community of fishermen.

THE GRANDDAUGHTER

O'Malley's is a group of on-the-beach cottages/cabins that includes a restaurant. Across the road (Norman Manley Boulevard), Uncle Oscar owns the best "jerk" carryout in Negril. The O'Malleys are a native Negrillian family. The natal home faces the road and the cottages/cabins were built one by one starting from the beach moving back toward the house. Those fourteen units were initially tent sites, where hippies camped out, serviced only by a pit latrine, standpipe-like showers, and no electricity. Now, most all of the cabins have electricity, and all of the cottages have their own bathrooms. Over the course of thirty years, the O'Malleys—father and mother—held the business together, building and expanding their enterprise. Failing health forced them to finally relinquish daily operations to their children, who by that time had already left Negril to live in Kingston or immigrated to the United States. Now, the management of O'Malley's is in the hands of Sybil, a granddaughter who grew up in Kingston but spent her holidays in Negril on the beach with her grandparents learning the business by performing family labor. Sent by her family to the United States to study business administration, she learned all of the advanced technological tools of the tourist industry. Sybil was given the task of "turning the business around." She must make O'Malley's competitive with other small establishments and maybe take on the new all-inclusive, RIU.

RIU, owned by a Spanish hotel conglomerate, is a massive hotel built at the far end of Negril. It is one of the new, all-inclusive properties boasting four hundred rooms that skirted the building code by erecting a hotel taller than the blight-ridden "Jamaica Tall" coconut tree. Not only that, RIU's daily all-inclusive rate of US$60.00 (2001 price) was the same charge as O'Malley's and other small hotels. Given a choice, a new Jamaica international visitor books a trip to Negril with agents in Europe or the United States and notices the cheap rate for RIU. More than likely, the client will stop there. Can O'Malley's compete against this global giant?

Sybil realized that RIU's "pre-fab" rooms (really just a series of concrete boxes affixed in a large structure) had little Caribbean charm, so she targeted certain sections for improving O'Malley's. Among the renovations were to secure the basics—reliable electricity, hot water, and available water during draught situations—and be consistent with room amenities, get cable television in all rooms, and join forces with other small cottage and hotel owners and develop their own website. Sybil envisioned a website that would feature "real Jamaican-style" not global "glitter" tourism. In addition, Sybil worked with the Chamber of Commerce in a similar venture for maximum exposure. Since O'Malley's hippie clientele is now middle-aged and returning with their children and grandchildren for vacation, their "word-of-mouth" promotional campaigns can no longer support O'Malley's survival as a domestic, native Negrillian operation. However, the website and the one organized by the director of the Chamber of Commerce, a woman at that time, focused on local and small business that would make a difference. In 2002, Sybil's job is not only to "turn the business around" but also to keep up the fight and help to ensure that Negril will not be swallowed up by the all-inclusive resorts owned by the Jamaican global giants—Sandals and Super Clubs, RIU, or the next foreign investor.

Next, *Women and Tourist Work in Jamaica* features an array of work experiences shared by women who participated in this study at various points in time and who work as chambermaids, now retitled as housekeepers. The narratives are organized by the size of the accommodation where these women work and their work environment. Since Negril is known for its unique laid-back and relaxing environment, then if one wanted a standard or five-star global tourist vacation, it is available too. Here, there are different kinds of face-to-face encounters with those who provide personal service and encounter vacationers.

HOTEL HOUSEKEEPERS

The following composite sketch provides a glimpse of the beginning of the workday for women who work as hotel housekeepers. Going to work, a woman housekeeper in a large hotel begins her day with about 5 a.m. by double-checking on her children as they ready themselves for school. She draws her sweater close to her body, picks up the bag that holds her lunch, and ushers the children out the door. All walk down to the coastal road, Route A-1, to wait for the bus for her and the van taking the kids to school. The worker joins a group of similarly hued women who are also waiting for their transportation to work in a yellow bus, a former U.S. school bus now owned by their employer, a resort complex on the beach. They see it barreling around the bend when it stops and additional women climb on board. In unison they say, "Good morning," to the driver and to the other passengers/coworkers. They are greeted by a similar response. As they take their seats, the coworkers chat among themselves as the bus makes several more stops along the way as it travels from Lucea to Negril. Another resort-owned bus travels from the opposite direction picking up hotel workers who live in the Sheffield area, a community on the Southeast side of Negril. The hotel provides this transportation for the majority of its workers—housekeeping staff—so they can get to work safely, in relative comfort, but more importantly on time. Once they arrive at the hotel property, they change into their uniforms, which invariably include an apron and cap. Reporting to the head of housekeeping, they get an update of the daily routine, and head off to clean guest rooms.

Discussions with three housekeepers featured here took place during their break time in the housekeepers' lunchroom, or when there was a prearranged time before they boarded the bus to take them back home. The first housekeeper tells a story about her working conditions and how she deals with those changes.

SMALL HOTEL

"Miss Ida" works in one of the older hotels in Negril. As times have changed, the clientele of this once-high-demand hotel has shifted accordingly. Now, Miss Ida reports, "respectable guests from Canada, Germany and even the nice Americans" go elsewhere. Now, she says, "we get people from Italy who don't know how to behave."

The Italy people who mostly young men come to Negril cause they hear about the nice beach, how things comfortable and so forth, and so on. Italy people dem also hear about ganga (marijuana) drugs, and rum. That's why they here. Let me tell you something. One day, I go to make up a room. You know we make up the bed, change the linen, clean the bathroom, tidy-up. I knock, even though da do not disturb is not on the door. No answer, so I let myself in. Well, there on the floor is this white Italy man, all dark and hairy and naked as they day he born. Empty rum bottles all over the place (Later on I learn that on the table were leftovers of lines of cocaine, but Miss Ida did not know what it was). The room is a wreck and it smells bad, bad. There's holes in the walls. The bathroom is filthy with vomit. My dear I tell you. I turn my heel go right out of the room to go tell the head of housekeeping the situation there. (Going back to the room), the man sits up, starts swearing in his language. He is showing himself to me and tells me that I better clean up this "shit" (now he is talking in English). Now, I don't have to take that out of order business from anyone. You understand? I look at the man, give him skin teeth and leave the room. Can you stand it? A drunk white man demanding of me? Massa, dem days is done. You hear?

MEDIUM-SIZED HOTEL

The realities of being in housekeeping in Negril bear consequences for the women whose job description is based on taking care of guest rooms, as well as addressing family obligations.

"Monica" works as a housekeeper for a hotel that features individual suites of three or more rooms. The work requires cleaning not only the bedrooms and bath but also the living areas, kitchen, and dining areas. She carries out her duties in the number of suites to which she is assigned. When guests arrive and enter their suite, she inquires what kind of cooking arrangements they require and if they have made them in advance (there is a beachside restaurant in the hotel and most guests have breakfast in their suite but have dinner where they can charge the bill to their room). If they plan on cooking in the suite, she makes her pitch. She offers to cook whatever meals they want for an extra price fixed by her, including menu planning, particularly for dinner. Monica will also shop for the guests and organizes her work schedule so that it fits with her culinary schedule. She might defrost a chicken while she is cleaning other suites and return to her "second job" as cook. Monica prepares the dinner and places it on the stove, so guests "just have to re-heat it whenever they wish" or come back to the room in the late afternoon and dine while the food is still hot. Monica gets high marks for her culinary skills, resulting in repeat visitors to this hotel because of these informal arrangements.

After they agree to hire Monica as their cook, she makes up a shopping list for them so she has the food items they like and those she can prepare. Jamaican-style food preparations are essential assets. Although Monica can arrange her schedule, when she is late doing any of these extra duties, she loses because she misses the hotel bus and must pay for the taxi to take her home. (I suggest she charge this extra expense to the guest; she said she would think about it.)

At home, her daughter, who does not have a job but is raising a toddler, lives with her parents and does all the domestic chores such as washing clothes, cleaning, and cooking. Monica's husband is involved in their church and has taken on this vocation since he was made "redundant" at the sugar estate with a very small pension. He also was disabled for a long period of time. When he was working, as a family, they built a house, and their children all attended junior secondary school. With little money coming in, heavy church tithes and responsibilities, the household wage-earning capacities fall onto Monica's shoulders. The extra money earned through her "informal cooking" is an essential part of their household income.

THE LARGE HOTEL

"Miss Mary Palmer" has been a chambermaid for a "very long time" at the same, now large, hotel. She has seen Negril grow from a handful of hotels to the two dozen or more now. Working at the same property accrues Miss Mary benefits, including being ranked by seniority among the household staff. Her daughter has a new baby, Miss Mary helps her extended family financially, too. A dark cinnamon-colored woman, Miss Mary is very plump. Miss Mary's father was a sugarcane cutter on a nearby estate, and her mother worked her garden plot. Like many working-class women of her generation, Miss Mary never went past the fifth grade. This is what she said:

I am one of the old-timers all right! We just watch the place grow from a few villas to now hundreds of rooms, C Block—that is all new. We staff see changes—managers come and go. Some of them are standoffish, some friendly. You know a good one when you see one though. I guess you can say that I train new girls on the job. I show them how to do things right, do it quick, quick, and make the guest dem feel comfortable. You have a certain number of rooms, and you wanna to be done before it is time to go, If the guests [are] from a foreign country, then they check out early, and you can finish the work.

My day? It start at daybreak. That is the old way—start at sunup. I do my chores, tidy up, get my lunch in the pail for me to carry, dress, and walk down the hill to the road. The hotel has a bus for the workers. Before the bus, we had to flag a van. Sometimes it come late or packed up with people. You get to work all crush up. When the kids were young, me, them, all had to be out at the same time. That was a time all right. Now, with just the baby girl and my grandson, I can take my time. We have to be ready to work, uniform on at 7 a.m.

On the third Friday [of the month] is staff meeting and we go over what is going right and what we must improve on. Last month, I was employee of the month. Thanks. Sure, I have gotten that award before, but this time the money was bit more, and I got a day off. I gave a little to my daughter for the baby, bought something for the house, and put some up.

What do I do? Let's say a guest is coming in. First, I check to see if everything is proper [such as] towels, bed linen—freshen up the room, make it nice. I like nice flowers in the little vases, not some dead-looking thing. In the villas I make sure that the kitchen things are in order, TV work. You know, the little things. When a villa guest wants dinner, I do that too. It costs extra. I ask them what they want, and the guest buys the provision, meat, fish. A truck comes every day and delivers to the chef and then comes around here. Guests can buy all their vegetables, ground provision, fruits from the truck. I buy (for myself for home) what I see I cannot get at the market. Guests can go the supermarket too, or the little store [on the hotel property], but the price too dear.

Most of the guests here are quite nice. Most give good tips and leave good comments. Some places I hear them nasty. Nobody wants to clean up behind that kinda business. We lucky here.

She goes on to say,

When it is slow, like it now, the staff is let go, starting with the new one, then go up. Long time ago, I used to be let go, but now I am one that stays on permanent. Tourism is good for Negril. It's getting bigger, but not as friendly like it used to be. Just as long as it don't get too big. [I asked her if a trade union ever tried to come here.] No, not that I remember. [I asked her if a trade union would help workers keep their jobs.] Not in this business, no; it too up and down. That is the bad side of the job, not knowing if management change, staff let go, hotel shut down; all that happen all of the time.

ALL-INCLUSIVE

Over the past decade, the development of large-scale all-inclusive resorts has rearranged the ethos and ambience of Negril, particularly on the beach. As British travel writer James Henderson (2013) commented, "The beach hums with those factories of modern tourism—the large all-inclusive hotel." There is little wonder that the physical size and the number of guests that each can accommodate does resemble a twenty-first-century version of industrialized tourism. Basic all-inclusive packages include lodging, drinks (alcoholic, no alcoholic), food (breakfast, lunch, and dinner and sometimes snacks), indoor and outside activities, and entertainment for a fixed price paid before arrival on the property. On point of arrival at the resort, guests are warmly greeted and welcomed to the property with assurances that this will be a fun-filled vacation, just as they imagined. This section features women who are on the professional staff of a few of the all-inclusive hotel in Negril. Each woman has advanced training in their respective jobs, and as members of middle management they look forward to advancement in their careers.

The following composite sketch describes, in part, the beginning of a professional woman's workday in Negril's tourism. Going to work at about 7 a.m., a set of women of varying shades of Blackness leave their homes headed for work. They make sure their work clothes hang properly so as to make a smart and professional image. Some don the uniform that marks their status as bank tellers, administrative staff, or agents. As they double-check to see that their children are readying themselves for school they see if a husband or partner or friend is giving them a lift to work. They wait for the honk of the car horn before going out of the door. Cars start down the street and turn onto Route A-1. They drive to hotels and assorted offices to prepare for the start of the day. Entering their places of work, this group of trained, professional women place their handbags in their desk drawers, arrange their work area, turn on the computer or other pieces of office equipment, and begin their workday.

THE FRONT DESK

"Margaret Bristol" has just come on duty as the clerk at the front of "Big Pink," a large family-oriented hotel. Margaret is working at night because she went out of rotation with the other clerks when she went on maternity leave five months ago. Actually Margaret notes that she likes night duty because it's very quiet, and luckily nothing catastrophic has happened on her watch. As her day begins at nine o'clock in the evening, Margaret joins other workers whose work hours are in reverse, night into day, the night shift. No matter the hour, the clerk at the front desk is critical to a well-run tourist establishment.

Being on the front desk is a low middle-management job. Margaret Bristol is well-prepared for almost anything to happen. She left high school with good grades and high scores on the Cambridge exams, but not high enough for entry into the university. Margaret did not take commercial courses but was able to easily pick up the routine of posting entries in lodge books and being an overall "gopher" in the "accounts" office at Big Pink. What she enjoys about working in a resort is meeting and having conversations with guests. Over time, she proved her merit in the hotel business and was given more responsibilities, changes in job status, and an increase in wages. This took four years before Margaret could become the night clerk. In that job, she posts late entries, usually from bar tabs or late dinners on guests' bills, answers questions, and checks in late arrivals.

At around midnight one uneventful evening, Margaret became comfortable enough to talk about the pluses and minuses of her job. On the positive side, Margaret was pleased with her job situation, and saw a bright future for herself and Big Pink in Negril's tourist industry. Big Pink had a great reputation with many repeat visitors. During the height of the tourist season, January through March, things really got hectic because there were "so many on property." In conversation, Margaret said, "The sheer numbers of people sometimes wears a person down when you are trying to help one person, one couple or one family at a time, and everybody is demanding something from you at the same time." Margaret addressed how she deals with the pressure of the job and simultaneously handles individuals and groups of people. She said,

> Well, I hate late arrivals, especially large groups, usually late with nothing to do with Jamaica. But something already set off the group, like a delayed take off because of weather, or worse something wrong with the plane. So, by the time they arrive in Montego Bay, everyone is annoyed. People working ground crew is fit to be tied. They have to wait until the last plane come in. The time they [the group of travelers] reach here, some have calmed down; at least they arrive safe and are on holiday. Others are just waiting for something else to go wrong. You know what I mean? They are arguing with their husband or wife, or friend, everyone around them. It's usually a man, makes himself like a spokesperson for the group and starts with the demands, even when everything is in order. One time, one guy was sheer rudeness, calling bad words, carrying on and cursing. There was just one bellman, and even though I asked a security guard to help handle the baggage, nothing was to please this man. He pushed in front of the line and demanded to see the manager cause of "inefficient staff" and "don't you people move under a crisis?" I told him that I was the night clerk and that he had to calm himself, and just relax [or] he would spoil his vacation. Real nice like. Well, that started off a whole heap of bad words and carrying on. I just kept on, took deep breaths, put on my best smile and got everyone registered, and assigned keys.

She continued to say: "You know, show skin-teeth and kept doing my work cause the man started to really annoy me to no end. You cannot let these people trouble you."

THE ACCOUNTANT

Mrs. Gwendolyn Hicks Brown is an accountant for a large resort in Negril. She began her work at the hotel as a bookkeeper and worked herself up the ladder. Gwendolyn "took accounts" (bookkeeping and accounting) in high school in MoBay (Montego Bay), as there was no secondary school in Negril. She went on to attend a commercial school and received a certificate in accounting. Married for three years, she has a four-month-old baby girl. A small, dark-brown woman, Gwendolyn is twenty-three years old. Just coming back to the job after maternity leave, she begins her workday at 8 p.m. Gwen remarked,

> I like working nights because you really don't have all the problems of the phone ringing, people asking you different things all of the time. Yes, with the

baby, I guess I am used to being up [in] the night. It's quiet; you can do the work and that's that. I come to work at eight, so I leave my house at about quarter past seven, twenty [minutes] after. By that time, I have given my daughter her feed [cornmeal porridge in a bottle], bathe her, dress her, and give her over to her father. We eat when my husband comes home at five. It seems a bit of a hurry, but he usually is hungry and I have time to cook. He loves the baby, so when I hand her over to him, they have a grand time together. He puts her to bed, then watches TV. The dishes? No, but he does put them in the sink. I do them when I get home in the morning. Sometimes I take the car and drive myself. Sometimes I get a lift. Sometimes my husband drives me to work.

Gwendolyn continued to describe her work:

When I arrive here, I retrieve the guest chits [register charges] that must be lodged, prepare the next day's bill for departing guests, etc. Sometimes like tonight, I am in charge of reception, too. There is a bookkeeper who does the actual figures—I used to do that job before I got my promotion.

Finally, Gwen says,

Tourist industry is very good for Jamaica, and, well, I should say for me. I always thought I would work for a hotel because in high school that was seen as the employer. I was lucky that I grew up in Negril area and saw Negril develop, so I got a job in the business. It was good timing.

As to her next job, Gwen replied,

Well, I don't know. If I go back and take more courses maybe I could try to go higher. It is hard after a point. Not here, at another hotel, maybe. Maybe here, who can say? Fees and books my family cannot afford now, and we have plans. I want to travel to Brooklyn. My sisters and brothers want to see the baby. Buy some things up there for the baby and for the house. So, I am saving for that trip.

THE TOUR DESK

"Diana Smythe-Jones" oversees the tour desk in one of the Negril's couples-only all-inclusive resorts. Her job involves promoting trips for guests to visit sites off the property, linking up with excursion operations, and making guests understand the cost and benefits of the trip (not included in the all-inclusive arrangement). She also reminds guests what to prepare for the trip, such as to wear appropriate clothing, shoes or sneakers, and other necessary equipment, and to recharge their smartphones, which at times lose power in the tropics. Posted around Diana's desk are posters announcing the off-property trips, times, and costs just to make things clear.

Diana was just getting off the phone with the excursion agent for Wet 'n' Wild cruise party. She had another set of guests in front of her who wanted a less wild event for their day trip. Diana got out the brochure for these two couples to consider. They huddled to confer with one another and agreed for Diana to arrange a lunchtime party cruise to Booby Cay (a small island off the coast just North of Negril), which was good for a short trip and great for snorkeling. Diana carefully explained when the couples needed to be at the resort's dock to be picked up by the cruise boat, what to wear, and when they would return to the resort's dock. The couples went off on their way for the business at hand, that is, fun in the sun and on the sand.

Two hours later, one woman from the group came up to Diana at her desk and asked for an itinerary again. Diana smiled, went over the agenda once more. Just as Diana was preparing to leave for the day, the same woman came up to her and asked the same set of questions. Diana thought perhaps there was something medically wrong with the guest. Everything she had said now twice was in the brochure, except for the times. Maybe that was the problem.

The next morning, Diana reported how the same woman asked exactly where they were to be picked up for the cruise. Diana remarked, "like I did not remember her and her pesky questions."

"I just show skin-teeth, cause now this woman is annoying me. I remind her once again. Do you know the same woman never said thank you all those times she bothered me, and then the last time she had nerve to be rude!! Like I should collect them from their room and escort them down to the dock because that is what I am being paid to do. Can you imagine? The rudeness of some people," an exacerbated Diana recounted.

SO MANY CHOICES OF PLACES TO STAY

There is no doubt that the range of accommodations available in Negril sets it apart from other North Coast tourist locations such as Montego Bay. When choosing where to stay in Negril, the pricing, location, and ambience are factored into the decision-making. No matter the location, there are women tourist workers who make it their business to provide the kinds of personal service required. Family cottage and small hotel owners and managers take pride in the efforts they make in providing a professional dimension to the hospitality offered in their establishments. Over the generations, the pride of ownership and the extra energy expended are due to it being a family business. Moreover, the stewardship of the property and hospitality are part of the management.

Woven throughout this discussion are the hidden gems of housekeeping, which are an essential function in the success of Jamaica's tourist industry. Each of these cases demonstrates the ways this achievement is accomplished and the efforts of Miss Mary Palmer, Miss Ida, and Miss Monica ensure that guests are well taken care of during their stay. In these cases, it was the sense of Jamaicaness that is presented and in the emotional labor meted out. The unfortunate incident shared by Miss Ida illustrated the downside of tourism as it can foster racism, sexism, and classism at its worst. However, Miss Ida took care of herself in a most Jamaican way—sucked her teeth—skin-teeth. Performances of Jamaicaness by both Miss Palmer and Monica yield repeat visitors and are also recognized by their employers. Monica's employer knows about her informal cooking even though it cuts into their restaurant business. These collective efforts exemplify ways of progress in the global tourist development schemes in Negril. Margaret and Gwendolyn represent the professionals of the tourist industry. Working at night allows them to see the negatives of the industry such as being meticulous in their work as well as working under pressure while handling tourist demands. Similarly, Diana also relied on her cultural practice of showing skin-teeth. The professional women use whatever is necessary to get them through the day and night and are generous in displaying their Jamaicaness and their confidence in the future of the industry.

Chapter 5

Entrepreneurs

"MAKE SOMETHING OF YOURSELF"

There are many business opportunities for women in tourism in Negril. Besides owning or managing family-owned lodgings or seeking career advancement in the industry, women are hard at work in "paradise." One of the goals of *Women and Tourist Work in Jamaica* was to illustrate the range of work that women are engaged in tourism. Not all of the possibilities are featured here, but the issues of differentials based on education, opportunity, and age are shown through the lives of the women who participated in the study. Of course, how, when, why, and where women are found hard at work is the topic of this chapter, which focuses on women who are entrepreneurs and own their own businesses. Those basic questions of when, why, and so on are framed by how strong the aspects of the gender system are in Negril, and if there has been any change over time. There are six locations that are these women's work sites: on the cliffs, on the side of the road, on the beach, in the sea, near the beach, and up the road. First, there is a historical component that is born by the example of the Jamaican higgler, whose contribution to society is both lauded and devalued. Gender, racial, and class stereotypes frame the discussions offered here and the points of view of the women who shared their stories in this study.

Much of Jamaica's gender system, as found in the West, was predicated on the dominant ideology whereby men were breadwinners and women were homemakers. The crack in this purview is how today's Black women, the descendants of the enslaved, are seen as astute businesswomen. Nonetheless, the contemporary small businesswomen, who are not higglers, are still tainted by the traditional image of market women, due to their aggressiveness. This mode of behavior is deemed smart and tough when attributed to a business-man. They may engage in other kinds of commercial and business services that are highly technical and capital intensive but remain in inequitable positions due to the general ideology regarding women in the economy (Carter and Cannon 1992). Of course, there are women who have been extremely successful in business, and Jamaica has quite a lot of them too. Each of those women's rise to the top included surpassing not only basic obstacles but also those that were gender related, such as proving way beyond reasonable doubt that they were worthy, despite being a female. Again, this perception relates to the contradictory nature of inequality inherent in the gender system.

HISTORICAL ROOTS OF JAMAICAN WOMEN ENTREPRENEURS

Trading has a long history in the Caribbean. West African women brought the concept of woman-as-trader to their enslavement in the Americas. In Jamaica, slaves grew enough food in house gardens to sell or barter some of it to other slaves and White masters. In vibrant Sunday markets, a West African tradition became a part of Jamaican culture—and stayed the domain of women. Over the next 130 years, noted Jamaican sociologist Elsie LeFranc (1989), "higglering," as it is called, changed little. This was the case of early settlements in Negril, when coconut oil and garden plot produce went to market in Savanna-la-Mar, or to stalls in Negril that the higglers staffed themselves. LeFranc argued the higglers' main goal is to be independent, as individuals and workers. A higgler determines her own timetable and marketing strategy, uses her own access to capital, makes personal decisions, and manages her own investments. Similarly, other small businesswomen, who may not be in the produce and microenterprise commerce arenas as the higglers, utilize the same goals, business strategies, and managerial skills. Some of these women learned their business skills in their family business, as a wife/partner paid or unpaid employee, and through their employment.

Entrepreneurship is valued in Jamaican society in the sense of "making something for yourself" as a personal goal. Further, the personal accomplishment aspect has proven to be a useful tactic if other qualifications are missing from a person's background. For example, "making something for yourself" can counterbalance the lack of middle-class or upper-class backgrounds, advanced degrees in education, and in the Jamaican context, light skin color. Entrepreneurship receives "high marks" because it promotes one of the fundamental features of "modernity" whereby individuals are ordained to achieve autonomy. In addition, "making something for yourself" assumes a masculine-gendered identity that is privileged given the prescriptive gender system.

In their study of a group of small businessmen in the UK, Scase and Goffee (1980) noted that the notion of the "self-made man" also had strong undercurrents in the UK. The same assessment is engendered in the United States. Further, there is a romantic element to this kind of entrepreneurship, including virtues of being enterprising, inventive, and generous with people while climbing up the ladder of success. Susanne Althoff (2020) conducted research with 100 women entrepreneurs in the United States. Common among those small businesswomen was the harsh reality of finding capital and finding clients for start-ups, even though they outperformed men in similar businesses. In addition, "Women seeking the kind of support men enjoy are stymied by the negative stereotypes and often sexual harassment" (Anderson 2020, N9). On the one hand, there are issues of lack of confidence, while on the other, too much assurance alienates investors. Overall, like their male counterparts, women share the same sense of drive, confidence, risk-taking, and goal orientation but are also motivated to be self-employed or evade other gendered obstacles put in place by employers.

Microenterprise development models are elements of the International Labor Organization's (ILO) core mandate for the promotion of social justice, the protection of women workers, and the promotion of equality between men and women in employment. Gender training in microenterprises affords women training, technical cooperation activities, and employment. For example, the Senegalese Federation of Women's Groups did not have the training to deal with contemporary financial practices that could increase their effectiveness and improve the lives of their membership. However, when two members who received microenterprise instruction sold their products, they reinvested the profits in another product, thereby enabling them to reinvest in both ventures and supporting their local credit union for the future investments.

In the case of India, the Popular Women's Micro-enterprises of Manipur focused on three of the most popular activities for women—poultry-raising, pig-raising, and weaving—and figured out ways that the poor women in the district could conduct this business, be close to home, and still perform their domestic duties. Access to informal credit was novel to this group, so they took advantage of quick turnaround profits from poultry-raising to reinvest in other rewarding ventures. The important factor here is that women's group's access to informal credit and savings was an innovative practice.

In Jamaica and elsewhere in the Caribbean, an informal credit system is called Partner or Sou Sou. This system came with the enslaved from West Africa when they crossed the Atlantic. A "banker" or treasurer forms a group who agree on a schedule of payment—weekly, biweekly, or monthly—whereby each member provides a "hand" or a payment to the collective pot. When a member meets the schedule, they withdraw their hand to use those funds for their intended purpose or can start again (reinvestment). A USAID project manager for the Microenterprise Project in Jamaica noted that the Partner system is widely used for gaining access to working capital for its poor and working-class contributors. In that report, a woman remarked that she did not like banks because that institution did not force her to save. She meant that banks do not have the obligations she feels she must fulfill with her Partner peers. Trust among the group and trust in the banker/treasurer are critical in the Partner system.

These sets of concepts, ideals, and practices were evident among the group of small businesswomen in Negril who owned cottages discussed in the previous chapter. The ten women whose daily life is presented here also follow similar senses of drive, risk-taking, and goal orientation toward their businesses, particularly since there is great competition for market shares in Negril. One case in point serves as a historical marker of the contemporary businesswoman—that is, she is a higgler.

"Miss Cilla," a "pet" name of Miss Priscilla (all are pseudonyms), is a dark-skinned woman with very little formal education. Her family has roots in Whitehall, a nearby community where they are considered poor, but stable. Her great-grandmother, grandmother, and mother grew subsistence food, were higglers by trade, and sold coconut oil like many in the old days of Negril. Now, Ms. Cilla represents herself as a visualization of a "classic" or "traditional" higgler. This representation is less about her economic activity but more about the visage for her customers. For her patrons who are tourists, she acts the part of the higgler, including her dress. Miss Cilla's business attire and activity are not quite a historical reenactment, but close enough. She calls herself a higgler, the tourists call her a higgler, and local Jamaicans perceive her as a higgler too. Miss Priscilla was able to skirt her working-class status and low levels of education because she was a part of the fabric of the community before the road was paved.

As the oldest participant in *Women and Tourist Work in Jamaica*, Miss Cilla's work life becomes the baseline for this chapter as well as a representative of the icon figure of mini-micro business, the higgler. As such, there is more to be said about the micro and small businesswomen featured in this chapter.

Following the daily routines of each of the ten businesswomen affords an understanding their lives in Negril. Further, as case studies, the narratives can be viewed as representative of others who are also engaged in small businesses. Through detailed descriptions, meanings of history, social hierarchy, and tourist development, these narratives illuminate a critical dimension of gender systems in a modern, global economy.

JACKIE'S ON THE REEF AND CLIFFS

In *The Pursuit of Happiness*, Bianca Williams (2018, 44) described the all-day lounging, oceanside massages, manicures, and pedicures of the interlocutors of her study at Jackie's on the Reef. "Girlfriends sometimes became lost in their thoughts and relaxing in solitude for hours." More importantly, this time at Jackie's was treasured by the ladies as many claimed they never took the time to spoil or treat themselves to these services, or to simply find time for reflection. Jackie (not a pseudonym), the owner of Jackie's on the Reef, has been in business since 1996 and her story is one of resilience. A former model and actress in Paris and an entrepreneur in Manhattan, she arrived in Negril with a dream of building a holistic spa. She spent a year living in a tent on the West End without electricity, sharing kindness and rainwater with her neighbors and living off her savings to build her dream on the cliffs. The main building is built of limestone that was quarried on the property. She offers massages and other healing treatments while clients listen to the waves breaking on the rocks close by. A swimming pool excavated out of the limestone overlooks the sea. There are guest rooms too. At the time of our conversation in 1998, Jackie was celebrating her successes, employing extra help, and expanding her services. In 2004, Hurricane Ivan almost destroyed Jackie's establishment. In an article appearing in *Essence* in 2005, Jackie says that "almost everything was destroyed." Forty-foot waves crashed right in front of her veranda while 140-mile-an-hour winds took down trees, roofs, and scattered debris. To rebuild her kitchen, message area, walkways, and assorted repairs, she used all her personal savings and funds from friends, who also helped with the cleanup. She reported, "I can always get beyond the problem and see the solution. . . . The devastation was filled with opportunity." Without the trees, a vista opened so that now she and her guests can watch the sun set over the water (Amber 2005). Fifteen years later, Jackie Lewis was featured in the *Negril Guide 2020* and recognized as a highly successful businesswoman who took the time to give back to the community where she lives and does business. As a founding member of Negril's Rotary Club, Jackie's motto is "Don't Let Fear Stop You."

MISS CILLA MAY: THE TRADITIONAL HIGGLER

"Miss Priscilla May," known as "Miss Cilla," is the most elderly of the group of ten entrepreneurs. Her long-term business experience as a higgler allows her to have tremendous pride as a vendor/market woman. Miss Cilla uses this "cultural and historical" element to her advantage in the way she conducts business. Miss Cilla represents "higgerling" as a true "Jamaica way" of making a living. Referring to some of the social factors of Jamaican society, her origins are addressed in her dark skin, low education achievement, and stable working-class background. Now, Ms. Cilla represents herself as a visualization of a "classic" or "traditional" higgler. This representation is less about what her economic activity entails but is directed toward her customers. Most of her patrons are tourists and her portrayal of the higgler includes her dress. Miss Cilla's business attire and activity are not quite an historical reenactment, but close enough. Miss Priscilla calls herself a higgler, and Jamaicans perceive her as a higgler too. Miss Priscilla was able to skirt such impediments such as low social status and low levels of education because she was a part of the fabric of the community before the road was paved.

Miss Cilla May is truly a classical higgler. She prides herself in being a higgler so much so that she is perhaps the most photographed fruit and vegetable vendor in Negril. In true higgler style, she sits on the side of the road behind a series of crates on which she displays the fruits and vegetables that she has grown herself, bartered for from her network of suppliers, or bought elsewhere. Beside her crates and displays is a gill cup, which is used to measure dry "peas" (beans), plastic bags, string, and newspaper. At a glance, she looks like a characterization of a higgler—big-breasted woman with a loose top, who wears a piece of traditional higgler cloth (plaid of red, black, and yellow colors) tied to her head, and sports a deep pocket apron over a long skirt. Miss Cilla wears sneakers without laces for comfort. Although her children are all grown and off to work, Miss Cilla usually has a toddler or two in tow who are her grandchildren. For our interview, she was by herself.

Miss Cilla learned her trade as a child as her own mother was a higgler. At that time, Negril produced coconut oil for cooking with its abundance of trees. However, due to the coconut tree blight, the trees died and so did the coconut oil processing and selling trade. As Negril's economy shifted from fishing and coconut oil to tourism, Miss Cilla branched out and included other provisions and fruits to replace the coconuts and the rendered oil. Tailoring her business with a new market strategy, she focused more on fruits than vegetables on a daily basis because tourists bought fruit quite steadily throughout the week. Her ground provisions and vegetables are sold to her local customers during the end of the week, when most of them do their marketing.

In the mornings, Miss Cilla checks the progress of her "paw paw" (papaya) trees, the ripeness of her bananas, and during the summer what mango, guineps, and avocado pear are ready to pick. She plants peas, yams, callaloo (a leafy green vegetable like spinach), and carrots. On her way to town, Miss Cilla stops by her neighbors to see if they have any fruits that she can buy from them, or if it is a Thursday, any ground provisions to augment her supply. She puts these goods in crocus sacks (burlap) and stands to wait for the bus that comes along to carry her to Negril. When they are in season, Miss Cilla will also buy pineapples in the large market in Sav-la-Mar, adding to her stock in Negril.

A beneficiary of Negril's expansion in the tourist industry, Miss Cilla's business requires that she computes the value of her stock and the pricing based on the demand for certain goods by different customers. For example, during a quiet time of the day, Miss Cilla puts a gill of "peas" in plastic bags that will be sold on the weekends. While she explains to a group of Canadians what a sour sop is, she slices one for them to sample. The major outlays of cash come from her buy and sell operations with her supplier neighbors and the cost of goods in the large market in Sav. She marks up these items according to points of purchase; for example, all bananas, whether they come from her yard or that of her neighbors, get the same high price. As a primary producer, Miss Cilla is dependent on good yields of crops and good weather. Another expense is the cost of traveling to Negril and to Savana-la-Mar. Although she does not pay rent, her spot for her crates on the side of the road is designated as her own by the court of popular custom. Vulnerability is when there is a major buy of a popular fruit; for example, someone buys all her stock of bananas or all of her guineps and she has to send customers elsewhere. She depends on her steady local customers for financial stability with her ground provisions because these are the people who will be there even if tourists do not come. Another major problem is competition when another higgler undersells her, despite the unspoken rule of respect among those in the business. Miss Cilla, who has a third-grade education, can calculate pricing, mark-ups, and the depreciation of stock in a flash without any technological support.

ON THE BEACH WITH MISS HORTENSE
MANNERS, THE PATTY SHACK

Miss Hortense never went further than sixth grade in school. A tall, thin woman, Miss "H's" mahogany-colored skin and "tall" (long hair) reflects her East Indian heritage. The patty shack, located on the middle of Negril's seven-mile sandy beach, is nestled between two hotels. The shack looks like a rundown tiny house, especially after five o'clock, when it is shuttered at the end of the business day. Miss "H" runs the business, although her common-law husband helps at the counter. Carrying insulated boxes on their shoulders, their sons hawk the patties along the beach on the weekends and during school holidays. During the winter months, the lines in front of the patty shack are long as hungry customers wait their turn to buy this cheap but succulent food. A patty is a delicious crescent-shaped meat pie made with highly seasoned minced meat, chicken, or vegetables in a flaky pastry shell like an empanada. Going on for ten years, the business started when her partner claimed a portion of the slip of family land that ran from the road to the beach. There was not enough footage to build a house or cottage, but just enough to erect the shack. The small wooden building has two doors, one leading out to the back of the edifice while the other is a side door where customers can file in, order, pay, and leave. There are no chairs or places to sit, but just the countertop. Ordered daily from Montego Bay, the patties are brought to Negril very early in the morning and kept warm in the metal oven behind the counter. There is a refrigerated case to keep drinks cold. There is no menu. Stepping up to order, just ask if the "Red Stripe is cold" or if there are veggie patties on hand, and that is it. When the stock of patties and coco bread (eaten with a patty for a major carb-loading effect) is depleted and her sons return with empty cases, then Miss "H" knows it is time to close shop for the day.

Miss "H" begins her day before the sun rises. Her home is located down the road from Negril in Orange Bay, and she does her domestic chores while her partner goes to Montego Bay. "Henry" leaves in his van traveling to the bakery to buy the day's supplies of patties. The patty shack is not the only distributer of this food item in Negril, so he must get on the queue to get his order early. Since this is a mid-July weekday, the order is medium (about 200 patties) with more beef than chicken and veggie. During the winter months and during "spring break" when Negril is full of U.S. college-aged tourists, the shack sells more veggie patties than during the rest of the year. Beef is the traditional filling, and a patty really is very much a Jamaican food. Unlike the bland British version, the filling of the Jamaican patty is spicy and most favorable. Most of all, the patty or coco bread is placed in a small brown paper bag with a juice box, soft drink, or beer, which makes a fine lunch. On the beach, Miss "H" sells the combination for US$4.00 (with a Red Stripe beer it costs $2.00 more—2002 prices); a fraction of what a lunch costs in a beach-side restaurant or bar. Miss "H" reaps a 25 percent profit margin. During the high season of the winter months, the sheer volume of business increases the profit, especially during the U.S. college "spring break" month of March.

Miss "H" has little overhead, except for electricity to heat the warming oven and the refrigerated box to keep the drinks cold. There is no lighting save for the sunlight that streams in from the doors. She does have to pay taxes to the members of the family that own that tiny strip of beach land. Miss "H" also pays her teenage sons a small wage when they work in the shack and hawk patties along the beach. Miss "H" works eight hours a day, six days a week, taking off only Sunday except during the high season.

On this mid-July day, the noontime temperature on the beach is "real" hot, and most of the vacationers are spending a lot of time in the water as they attempt to cool off. The temperature inside of the shack is suffocating, but standing behind the counter filling orders is Miss "H." She is sweating in the heat as she hands over a small paper bag with a patty, retrieves an orange box drink out of the cold case, and collects the money from a visitor. Since no one wants to linger in the heat, customers file in and move out quickly. Today, there are a lot of drinks being sold as well as the patties. Pleasantries are expressed, all having to do with the weather. The sons venture out onto the beach, selling to tourists right in their beach chairs, or at the water's edge. The patties are already in the small paper bags and the "traveling" metal box that holds them is soon empty thanks to good sales. With two points of purchase, customers do not have to walk over the hot sand to buy patties and a drink. As the day progresses, Miss "H" calls out to her partner about the size of tomorrow's patty order and checks to see if the drink supply must be restocked due to customer demand. At 5 o'clock, Miss "H" closes up the shack and the business day is done.

On another day in another year, the beach is the site of this research, but the workspace is underwater.

CAROL MARTIN IS UNDERWATER

"Carol Martin" owns and operates a dive shop in Negril. At one time, it was only an independent dive shop in Negril, and Carol bought the business from the original owner in the late 1980s. Carol came to Jamaica as a tourist in the early 1970s. Drawn to Negril for its laid-back ambience and non-pretentious welcome, Carol, a young White American woman, is now Jamaican by residency and marriage. Besides falling in love, Carol also fell in love with the village with its beautiful seven miles of sandy beach. But more importantly, she fell in love with the reefs, coral, and underwater wonders. When Carol decided that she was not going "back to the States" but would try to make a life in Jamaica, she found that she had a needed skill. Carol was a certified diver. Through work as a diving instructor and as an underwater tour guide, Carol knows almost every nook and cranny under the water's surface. There are over 1,000 species of coral, fish, sponges, and other fauna in the nearshore waters. Over the years, Carol Martin has been part of the group conservationists who have organized to protect the reefs from their major predator—tourists. Inexperienced fins and hands brush against, knock over, and step on fragile coral species. What the tourists do not damage, non–ecologically minded fisherman and divers foraging for black coral to sell at high prices to visitors continue to harm the reefs. Of course, this all happens on top of pollution, oil spills from boats, and poorly designed drainage systems.

When Carol bought the business, her mission was to make a living at doing what she loved to do, SCUBA dive, and to do it well. Over the years, Carol has certified most of the Jamaican-diving instructors in the area and schooled them in aquatic ecology. Her goal was to certify Jamaican women as diving instructors and underwater guides. The staff includes mostly women, but none are donned in mask, fins, and tanks.

Each day begins with checking the sign-up sheet to see what activity is planned and who is assigned to do that work. The dive shop is a business that uses expensive equipment that must be checked for reliability. People's lives depend on it. After a recheck of equipment, tanks refilled, boats gassed up, flags and emergency equipment checked over, the staff awaits the first group. On this day there is a morning session of intermediate divers who are looking for a fun dive, while the afternoon group wants an adventure and will explore one of the wrecks off Booby Bay, just outside of town. Carol goes over the staff's work schedule, looks at catalogs for replacement equipment, takes at least three telephone calls, and makes another one before the 10 a.m. group arrives. Her clientele is self-selected because they carry PADI certification or go through her program. However, she will not allow even paying customers to use her services if they do not meet certain qualifications of responsibility and understand the serious nature of the sport. Over the years, business has been steady but follows the cycles of any subsector of tourism.

Besides selling lessons and guided tours, the dive shop also sells snorkels and masks, and Carol does not dive anymore unless there is an emergency or she wants to demonstrate the effects of her conservation. On this day, she escorts a party of one person to a dive site that is her pride and joy. This area had shown signs of reef devastation. Beckoning with hands, she points to the plants that are coming back to life as well as to the fish. The coral will take years to redevelop, but now their growth cycle will not be interrupted by careless arms and out-of-control fins. Swimming with ease and grace, Carol is at home under the sea, the source of her livelihood. Carol is a businesswoman whose care of the reef and marine environment is augmented by her activism in preservation efforts in Negril's waters.

Back on the beach but moving away from the waters' edge are entrepreneurs who focus on hair braiding and hairstyling. Their workplace is located under umbrellas, perched under a tree, or inside a resort's salon. Topics that surfaced here focus were why hair braiding is a tourist enterprise, who are the women engaged in this type of business, and what elements of service, hospitality, and professionalism enter this line of work.

HAIR IN NEGRIL'S TOURIST INDUSTRY

One of the most defining symbols of femininity is hair. Even Oprah Winfrey (1997) recognized that "no matter what your race or where you're from, in a society that makes judgments about people based on what they look like, hair is a big part of how you look" (1). In Jamaica, a source of inequality on multiple levels of social difference is the texture of hair. Nappy, kinky, dark, thick hair that stands still in the wind harks back to its African origins, which up until the later part of the twentieth century, were looked down upon in terms of class and color. Looking back to the period of enslavement, it was hair, not skin color per se, that became the more potent mark that symbolized servitude, and "hair type rapidly became the real symbolic badge of slavery communities" (Banks 2000). These are the issues of class, race, and color categories that were constructed and maintained by those in power with the force of the dominant group.

The legacy of enslavement engineered a society where race/color, gender, and class were critical components of social standing. For women, physical features close to the hegemonic beauty ideal were looked upon more favorably than others, although the majority of them have dark skin. The British colonial social order continued through independence in 1962. In the 1970s, when light-skinned, brown elite Michael Manley was elected prime minister due in part to his dark-skinned, Afro-wearing wife Beverly, there was acceptance of "bushy" hair and Afrocentric hairstyles as part of the nationalist creed to embrace the African side of the Jamaican motto, "out of many, one people." Even then, a preference for light skin color (brown, light brown, tan) and straight hair was the rule of the day for women to be considered feminine and beautiful based on hair texture and skin color. Women with long/tall and wavy hair were prized for their "femininity" and "pretty hair." For the majority, hair that was straightened by a hot comb or through chemical hair relaxers was the only way a dark-skinned girl could overcome the skin/hair bias, even if she was a member of the middle class. If she was poor and working class, the road to social acceptance required the completion of secondary education, and the more advanced training the better if she was to succeed in her goal of upward mobility. Since hair relaxers were imported from the United States, poor women stretched the timing of their "touch ups" when the chemicals were applied to new hair growth. They did this hair maintenance themselves or with a help of a friend or family member. The difference in hair management and upkeep became a class marker, often signaled by abused and overprocessed hair.

When the experiment with radical politics ended in 1980 with the election of the conservative Jamaica Labour Party, and Rastafarianism influenced only a segment of the educated elite and portions of the working class, the embrace of Africanisms in hairstyle waned. Jheri curls, chemical relaxers (crème), hair weaves and other hair care products, and styles that produced straightened or non-kinky hair were prominent again with the same class distinctions Jamaica. "Good" hair looks implied upper mobility if donned by a dark-skinned woman because they could afford to pay for this hegemonic beauty symbol. In contemporary Jamaica, the range of hairstyles worn by women and girls run the gamut from dreadlocks to braids, sister locks, Senegalese twists, weaves, cane rows, and more. Afrocentric styles are tolerated in situational contexts. How women wear their hair is based on their socioeconomic status and skin color, plus the self-confidence they wield to do so. As late as the 1990s, school-aged girls were still discouraged from wearing Afros or plaits in Afrocentric styles in private and government schools. Therefore, Jamaican tourist workers/hairdressers whose own hair communicates their cultural practices are tested by the visitor (European, Euro American, Euro-Canadian, and sometimes African American).

For the tourist industry, hairdressing and braiding is a specialized segment of the services provided within the wide range of the tourist business—from high end to mass tourism in the middle to those on a very tight budget. According to academic and professional literature on hairdressing, *trust* is the operative word. The element of trust is tested when the woman and her hair go on a vacation to Negril. Depending on the wishes of the client, how do poor and working-class and skilled professional Black Jamaican women gain the "trust" of their visitor client in styling their hair?

Hairdressing is a personal service involving a face-to-face encounter. Featured here are three groups of hairdressers/hair braiders in Negril whose narratives illustrate the nature of those face-to-face encounters. The first is the itinerate hair braider who trods up and down the beach looking for clients all the while dodging the security staff, whose job is to not let them on their guarded piece of the beach. A second group of hairstylists are hair braiders who are available near small- and medium-sized hotels. If they are associated with the small hotel or guest house, the hair braider will have certain rights and privileges, such as a designated spot where they set up their chairs and tools of their trade.

Early on in this research, the Negril Chamber of Commerce licensed a group of women as hair braiders. By paying an annual fee, holding a photo ID, they also attended courses on small business management and cleanliness organized by a U.S. Peace Corp volunteer. Located in their own kiosk booths, these licensed hair braiders sometimes wore a pink-colored uniform as a sign of their position. After ten years or so, when those kiosks were in disrepair and subsequently demolished, a new plan was put in place. The licensed hair braiders, now accompanied by massage workers are now found on hotel beaches and are members of those establishments.

Finally, there is the professional beautician ensconced in beauty shops in every major hotel in Negril. In the high-end properties, these professionals also provide a variety of skin, nail, and hair services. If not on the hotel property where a client is a guest, a hairdresser can be called in for special events, such as for tourist weddings. The skill levels between the itinerate and licensed hair braiders may be the same, but there are other markers that make a difference among the group as well as to the tourist. The observations and narratives of these women workers show if and how trust was established between them and their vacationing clients. Also entertained are matters such as this: How do licensed hair braiders and those who braid hair on the beach deal with their vacationing, silky-haired clientele? For the professional hair stylist, when does a woman "trust" her Jamaican hairdresser/braider to keep the same look, or does she change the look for something adventurous, like her trip to Negril itself?

Observations of hairdressers and hair braiders at work in Negril relied on their availability to carry on conversations at their workspace and if they had the time to do so. There was a bit of participation done to strike up a verbal exchange by all three sets of women hairdressers. All the women set the price of their service according to their location. The beach braiders who walk up and down the beach offering their services to potential clients set their own price depending on the number of clients serviced. In the morning, the price may vary to what happens over the course of the day. Licensed hair braiders are found in specific spots near a small hotel, beach cottage, or restaurant, where they have a business by kinship. They pay their "host" a fee for the space depending on the connection; the price of their service depends on the time it takes to accomplish the task and is set by the membership of the Licensed Hair Braiders Association, a group established by the Negril Chamber of Commerce. For those who work in resort shops, the fee is set by the hotel management for a range of services—cut, blow dry, styling, and so on. They also can braid hair as a "style." The following are samples of hair work in Negril that begin on the beach.

Angela is what the Chamber of Commerce calls a scourge on the tourist industry, an itinerate aggressive vendor who "harasses tourists." As a beach braider, she walks up and down the beach, usually starting down near the bottom of the seven-mile Negril beach and going as far as she must to find enough women and girls who want their hair braided so she can make the day's work worthwhile. She does not know who started the popularity of White women wearing cane rows in their hair, but it has provided a way for her to make a living, as meager as it is. The twenty-nine-year-old Angela left school after sixth grade because her family could not afford to continue her education. She took care of her siblings and performed the family household chores. Soon boys called on her, and one thing led to another, and she was a mother. For a while Angela did days' work at a bar (cleaning the restrooms, washing and sweeping floors), but that ended when one of the bartenders tried to "make a move" on her (sexual harassment). Angela says she cannot be bothered by that kind of thing. Braiding was easy enough to do as a moneymaker because she already knew how to plait hair well. She plaited her sister's hair as well as her own for as long as she could remember. At present, her own hair shows signs of being over-processed and damaged due to the harsh chemicals used in the "crème" to straighten her tight curls. In terms of her clientele, the thin hair of White tourists took time for her to learn how to manage, but she has become skillful. As she walks the beach, Angela carries the tools of the trade with her: combs, beads, a mirror, and a folding chair for the client to sit in. During "Spring Break" when thousands of young Americans come to Negril for their holiday, Angela makes "good money." That four- to six-week period is the major source of her annual income. The Spring Breaker girls, hung over from too much partying the night before, are compliant customers as they sit in her chair, ignoring dirt-encrusted combs and Angela's unwashed hands. The element of trust is not an issue here, but the adventure is ever present. Angela passes pleasantries with these college students but does not understand what they say in response to her or to their girlfriends who lay on the sand nearby. The girls do not want a complete outdated "Bo Derek" look, but just around the edges of their hair. Angela charges them US$20.00 (2001 US$26.68) each for the service, and she explains that this extra $6.68 dollars is less than the "expensive" licensed hair braiders whom she sees as "too thief."

Cherry is one of those "too thief" licensed hair braiders that Angela sees as her competition. In fact, the competition from the unlicensed braiders is a sore point for the members of the Hair Braiders Association because of the large numbers of women who do similar work. However, there are major differences between beach-trolling braiders and Cherry's group, such as working conditions, repertoire of styles, and tools of the trade. With the help of the Chamber of Commerce, Cherry and two other licensed hair braiders and a woman who offers massages built a kiosk to house their services. Open on all sides, chairs in the kiosk look out onto the beach in sight of the area where the table of the licensed massage "therapist" is located enclosed by a canvas tarp for privacy. Like Angela, Cherry had to leave school early but finished junior secondary school. Cherry wanted to become a nurse, but the money was not available for her to continue her studies. All Jamaican girls learn to plait, and Cherry was no different in learning this skill. Grooming was a part of the education she received in her home in nearby Orange Bay and hair care was part of the routine. In school, girls wore their hair in plaits. As young girls when you got older in age, "you begged your mother or received as a gift your first 'crème' [relaxer]." The upkeep process for relaxed hair was in line with the economic fortunes of the household. In her teens, Cherry looked for jobs that she thought to be suitable for the goals she set for her life. A Peace Corps volunteer assigned to the Negril Chamber Commerce provided classes for small business owners and those with entrepreneurial potential. Cherry signed up for the classes. One of the new arenas in the tourist industry was the licensing of hair-braiding practices. Beach braiders were seen as a harassment and a deterrent to the industry. Some women on vacation did not want to be hassled about having their hair braided, among the many other products vended on the beach. The Chamber set up a scheme that was attractive for young women like Cherry. They could hone their entrepreneurial talents, while learning multiple styles of hair braiding. In addition, the course exposed Cherry to the basics of hygienic practices expected by tourists and to understand the benefits of being a member of an association.

Cherry and her friends take turns as to who opens and who closes the kiosk. They share the costs of the space, its upkeep, and the replacement of tools of the trade. They wear pink uniforms that symbolize their association and their membership in the Negril Chamber of Commerce. All the women look neat and tidy, a way of making tourists comfortable. One of women of Jamaican-East Indian heritage has long but very curly hair. Cherry and the other braider and the masseuse have relaxed hair. Situated between two medium-sized beachfront hotels, Cherry's location provides a steady business. When one of hair braiders does not have a client, she helps out the other, which makes the task go quickly. This is especially the case for clients who want all their hair braided with many beads, popularized at the time (1999) by tennis players Venus and Serena Williams. As they stroll along the beach, Spring Breakers stop by the kiosk and ask the price. Some balk at the US$60.00, but others rely on the attractiveness of the place and perhaps the connection with the hotel where they are staying. It is a sign of trust.

For African American clients, Cherry says they come to her when they stay in Negril because she reminds them of their stylist at home. Unlike the Spring Breakers, who are hesitant at the price and turn to trolling beach braiders, African American tourists assume a sense of trust that comes with the extra price. By 2018, Cherry wants to learn the new technique of doing "sista locks" because she heard that the style is popular in the States as well as among affluent Jamaicans in Kingston. To do "sista locks" requires the use of an instrument like a crochet hook that winds around a small section of the hair, securing it and locking it. With thick, kinky hair, sectioning hair takes hours and hours. Unlike individual braids that might need synthetic hair to produce a certain style, no extra hair is needed for "sista locks," just a lot of time. Cherry thinks if she can be trained to do "sista locks," she might add it to her style list and maybe work in one of the mid-range resorts and attract an African American clientele. She wants to pursue a contractual basis with a hotel to do this hairstyle.

On the top tier of hairstylists in Negril are those who are professionally trained. Born and raised in Kingston, Lucene is a beautician who looks fashionable with relaxed hair, makeup, and clothing modeled after whatever is in vogue in New York or Milan. Lucene went to cosmetology school in Kingston after leaving school after fourth form (tenth grade). In cosmetology school, she learned all the elements of hair care, hairdressing, and styling for different textures of hair to earn her certificate. For a couple of years, she worked as a stylist in a full-service salon in Manor Park in uptown Kingston.

Lucene left her job in Kingston when she answered an ad posted in *The Sunday Gleaner*. Her notion was if she did not like Negril or the job (subsided salon booth rental), she could always leave. Ultimately, her hope is to save enough money so she can immigrate to the United States or Canada as she has family in both places.

Lucene's job has two speeds—slow with little or no customers or fast and out of control when there are wedding parties. Lucene is often bored at work, especially when there are no special events, like a large wedding to service. For the most part, many of the women vacationers at this resort have their hair styled before going on their trip. Breaking the monotony, Lucene sits in the shop, reads magazines, and often does her own hair. To keep up with the latest fashions and hair colors, Lucene explores the trends in both dominant culture and Black beauty markets.

One wedding stands out in her mind because there were several women to service all at one time. The couple was from Missouri. They planned to marry in Negril, where they met five years before, ironically during Spring Break. The resort wedding package was ideal, and friends and family flew to Jamaica and Negril for the event. The size of the wedding party was the conundrum, as there was just Lucene by herself. The resort had staff that could help her by doing manicures, pedicures, and shampoo for the party. All of this meant there was styling, primping, and making perfect the bride, her mother, the mother-in-law-to-be, the matron of honor, six bridesmaids, and the one flower girl. Lucene's organizational skills saved the day with the plus of the extra staff to do the unskilled labor, such as providing shampoos for ten women and one girl. The resorts' wedding planner worked with Lucene in terms of what the couple was looking for in the overall presentation and the color scheme. Using local flowers for the flower girl, the matron of honor, and the bridesmaids, the uniform style was changed only by the length of an individual's hair length. Upsweep hairdos were done for the mothers. To ward off the humidity, gel and hairspray were liberally used. As is usually the case, the bride wanted something different. Thinking about the hairstyle she had worn on her first visit to Negril when she met her fiancé, she wanted to recreate that affect. To do so, Lucene braided flowers into the bride's just-below-shoulder-length hair that echoed the braids and beads of that earlier hairdo. The bride was the last one to be finished. Working almost eight hours nonstop, Lucene checked every woman to make sure they looked perfect and then hugged the bride. Not only was Lucene a trusted service provider, but she was also a most capable one too. A bonus was the very big tip from the father of the bride who paid for the salon rental.

Hairdressing, like many other services, varies in skill level and is augmented by social indicators of race, color, and class. One who makes you "feel good all over" as the Jamaican Tourist Board slogan proclaimed. Of importance here was how hair represented a beauty standard for one group that was manipulated by another to achieve that same sense of splendor and femininity. How hair is defined and commoditized by the hairdressers and hair braiders in Negril represents just one aspect of the equation here. For peoples of African descent, hair matters for good grooming, but also as a way of becoming "civilized" according to the ideologies emanating from hundreds of years of British colonialism. All the women had straightened their hair to be presentable to tourists and to themselves. How well that process was done depended on their economic situation. Skin color was constant; the symbol of "good hair" was critical.

Does the service provider's own hair have anything to with their performance on the job? The women represented here show two things. First, if you are poor and Black providing a service, that makes you basically invisible to the client. The Spring Breakers made little effort to engage in conversation with Angela and she did not care for their conversation anyway. Her trust and proficiency in braiding their hair was not in the equation. Angela took advantage of the overseas students' adventurous state, charged for this experience because she was the only low-cost opportunity for them, and then was able to make a living for herself. The NCC-provided kiosk allowed for visibility for Cherry and her friends, and elements of trust do come into play. Vacationing clients make choices as to whom they turn to for this service and want a provider they can trust. Although dependent on the health of the tourist industry, Cherry's steady business benefits from the trust she has with the nearby hotel, which directs those tourists to her kiosk, and later on in front of the hotel. Of course, Lucene is most trusted by the client because she is a professional on salary who comes highly recommended by the resort. And what do these three women think about their relationship with the tourists? Angela, Cherry, and Lucene echo similar sentiments. It might be an adventure, but when the visitor returns to the United States, Canada, Germany, or wherever, all will know that they have been on Caribbean vacation because of their hairstyle. They will tell all their friends about the beach, and some of them will come to Negril by this word-of-mouth endorsement.

UP AND ON THE OTHER SIDE OF THE BEACH ROAD

On the other side of the Beach Road is a craft market whose clientele are directed to it by fellow members of the Negril Chamber of Commerce as well as those who just venture into this location. In addition, individual vendors make once-a-week scheduled stops to a conglomerate resort. They carry carefully selected inventory with them and set up their displays on resort-provided tables. No matter what their location at the market or in the resort, an act of resistance is applied by these women vendors, who call out to "make me an offer." "Make an offer" is a cultural expression that shows disdain for the budget-minded tourist (usually of U.S. origin) who shows a lack of respect of the vendors, their workplace, and their personhood. Of importance here is that each group holds each other in total disregard and each maneuvers their responses toward one another as they seal the deal to "make an offer."

Located in a government/craft vendor association market located across from the beach, the craft market is where tourists and tourist workers play the games of cats and mice. One's survival depends on how well they outmaneuver the other. Tourists are sent to the crafts market following the advice of a hotel worker, carried there by a taxi man, or they find the spot while looking at the location on a tourist map. The tourist enters the market with a sense of adventure, captivated by the bright colors of the garments blowing in the wind. The stalls, made of bamboo and sideboard wood, are filled with stock ready for "gamesmanship." Lured to craft markets by the necessity of buying gifts for those at home, tourists also want to purchase items that fit their wallet/pocketbook. There is even a T-shirt that voices that state of mind and economy: "Grandma went to Jamaica and all I got was this lousy T-shirt." Vendors come out of their shop to greet the visitors by imploring each to "come inside and see for yourself. If you like something, me make an offer." For most U.S. visitors, prices are usually fixed at a price point. However, given the heads up, tourists know that in Jamaica and particularly in a crafts market, they must "haggle" to get the best price and are expected to do so. It is assumed that after the initial price is stated, a very low bid would be welcomed. Contrary to that opinion, that very low price is deemed as a point of disrespect on behalf of the vendor. Why should a vendor "give her stock away"? From the point of view of the vendor, here is another cheap American, probably someone on a budget, and by their actions views them with contempt. This could be an issue of class as working in craft market is not acknowledged as a respected occupation unless it comes with other accouterments. It could also be a national issue, a race issue, or a gender issue, or one that intersects all of those possibilities.

The following was shared about the situation. The president of the craft market association, a woman vendor of great repute by her peers and community, remarked about national identities of customers and notions of respect. She said, "Dem come on a cheap tour, they work hard to get the money to come to Jamaica, but n they look pon we and say all a we must come cheap too. Dem poor Americans want is dear for nudding an dem think we foolish and sell dem fe nuddin." A case in point are the straw baskets that are sold usually with *Jamaica* or *Negril* written in raffia on the sides. A traditional tourist item, straw work can have special notations, such as names, dates, and so on, that can be customized for an additional price. A crafts vendor who does that kind of work remarked that it takes time to do the needlework and that the budget customer is clueless to this factor. She noted that all they see is straw and it must cost nothing in terms of time and energy let alone materials. The reaction of the crafts vendor is to "hard ball" the sale—meaning to make the sale, they must "make an offer and get on with it," even though they are grinning and bearing it showing skin-teeth. For those who are hustling and harassing tourists who are seen as "moneybags," and those who are in legitimate business ventures such as craft marketing, all recognize the difference of "who is who," much to the chagrin of the visitor. One U.S. visitor was overheard asserting that the Jamaican crafts women really acted as if they were a special gift from God because they did not "kow tow" to his demands. What he did not understand was that the "make an offer" was not an arrangement whereby he had the upper hand but was an act of resistance against the world where the customer is always right. "Make an offer" was loaded with hidden innuendos of disrespect on the part of the Jamaican woman craft vendor as illustrated by the interlocutors.

"Mrs. Green" makes specially made handcrafted dolls and as an elderly woman arranges her display of goods according to the range of price points in her shop, targeting specific consumers. After years of experience, Mrs. Green can assess a customer as soon as they walk into her place. The dolls are displayed in a way that designates an attractiveness factor for a set of visitors. For example, she directs Americans and Italians toward one area that she assesses as cheap, or poor-quality dolls. Germans, Scandinavians, and Japanese customers are pointed toward the high end of the stock, with Brits and Latin Americans located in the middle area of the craft cottage. Mrs. Green guides those customers to a particular shelf with tremendous accuracy while anticipating their preferences. When she detects a wealthy American, she is rewarded for her efforts when she steers them to the "high-end" shelf. A customer who recognizes the quality of the craftsmanship and spends time with Mrs. Green often finds that she provides a different kind of offer than the tourist "make an offer" usually signals. Mrs. Green incorporates her sense of Jamaicaness that is a sign of mutual respect and is afforded to a respectful visitor. These are the stories the Jamaican Tourist Board wants to promote, but it is impossible to do so because Jamaicaness must be warranted by actions and deeds on behalf of the tourist and not the other way around.

At one of the two craft markets in Negril, "Penny Lewis" has sold T-shirts, straw crafts, and beads for the past ten years. She enjoys the work and also takes turns at being the president of the craft market association. At the time of the research, Penny took advantage of the management course offered by the Chamber of Commerce. The classes were run by a U.S. Peace corps volunteer interested in small business. Tall, slim, dark-brown-skinned, with a flashing smile, Penny, at age 34, has eight years of schooling and is a single mother of three children aged 16, 12, and 5. She recounted her daily experiences:

> I started the day off early as usual, about 5 a.m. In the old days, my mother, she didn't have no clock, just the sun come up and she start about her business—she work, the land. This morning, after I got my daughters up and the boy, go to school, we had a little breakfast, made sure there was lunch money for them, and then they went off. Tidy up the place, and then I look at my list again and go over my money. I decided last night that I need to get a new style T-shirt and more colors for the straw [raffia]. What I've a do is go to MoBay to the factory for the T-shirt . . . go look for the goods at the shop.

She went on further to say,

> So, here it is morning time and I nyah go to the shop to make money, I go to town and spend money! That takes all morning and way after noon, till what is it almost time to close up the shop? But I need those new things to keep up, so the trip was okay. Tourists like new styles, especially them that repeaters. See this shirt? It costs me more than this one here but is feels nice-nice and soft. Now I could not make this trip in high season, unless somebody I trust run the shop, they don know how the price go. We keep the same here [referring to the craft market association members], but sometime do our own thing for a special customer. You know what I mean? But now, what is it November? I can take the time.

Finishing up the conversation, she said,

> When I leave here, it's late now. I take the van home to Orange Bay. My daughter starts the dinner when she gets in from school. It's a help. The boy does [washes] his own uniform now, too. I finish the dinner. We eat. The kids do their homework. After tidying up the house, I set on the verandah and work on the straw. You know, put things on the basket like "Jamaica," "Negril," "One Love," tings like that and flowers, More details cost more to make and make a high price, too. I work a bit, listen to a program on the radio, then lock up for the night. Sometimes somebody comes to call, but tonight, I too tired.

MAKING SOMETHING OF YOURSELF

The details of how, where, and why women in business in Negril do well in the tourist industry are clearly demonstrated by the experiences of Jackie, Miss Cilla, Miss H, Carol, Angela, Cherry, Lucerne, Mrs. Green, and Penny. Collectively, they all show how tourist-related activities fall in line with what the local community hoped for in the development of tourism. Families like that of Miss Cilla were able to make a decent living from the tourist trade, despite not owning land or having an education. Jackie, who owns the land and the business, was a champion who fought the elements during Hurricane Ivan. Miss "H" does not own the land her patty shack occupies, but it is such a tiny slice that her paying the taxes to her partner's family and a monetary contribution in the form of rent makes it all worthwhile. Naturally, the dive shop is ideal example of business and environmental preservation, and Carol led the way in education and activism in terms of the reclamation of some of the damaged reefs of Negril. The personal services provided by the hairdressers show how their businesses provided fond memories of those experiences for visitors through their businesses. Each of the ten had a distinct clientele and they serviced them well.

As each narrative described, in the face-to-face encounters between them and the visitors, they performed a service with varying levels of mediation, business acumen, and resolve using Jamaican cultural practices of skin-teeth and Jamaicaness. Invisibility and disrespect is on the one hand, with respect, trust, and valuation on the other; these are basic indicators of what is tolerated and embraced as a benefit of being an entrepreneur.

Tourism in Negril relies on all of these women for its image, service, and accountability. It is on that very local level of interaction and business transactions that businesswomen make their contributions to the success of Jamaica's tourist industry, make a living for themselves and for their families, and for some to make a profit. Another illustration is the scope of the kind of small businesses in which Jamaican by choice, Jamaican native born, or Jamaican by marriage women are engaged. In tourism, women take advantage of this range of options according to their access to capital, training, and education. Whether they recognized the historical fact or not, this is Negril's part of the Jamaican legacy of the savvy, aggressive, shrewd businesswoman, also known as a higgler and an entrepreneur.

Chapter 6

Nightlife

Nightlife focuses on the "workday" of the women tourist workers in this study whose work takes place during sunset, in early evening, and/or into the night. They are all involved in food, beverage, and entertainment arenas that are essential components in the tourist industry. Since 1970, the increase in the number of entertainment venues illustrates this point. More than ever before, provisioned by the steady hands of a woman worker, tourists are offered up numerous spots for their relaxation and enjoyment.

Besides the all-inclusive resorts, which have their own in-house entertainment, there are varieties of places eat, to hear music, and socialize that extend all along A-1/Norman Manley Boulevard/Beach Road and up alongside the cliffs of the West End. This bounty of entertainment was the reason why many tourists come to Negril in the first place. Geographically situated facing west, Negril's fiery sunsets are dramatic to say the least. Many beachfront establishments accommodate visitors with the best view possible of this daily orange ball dipping into the sea as they consume a beverage of choice and snacks at their table or at the bar. An adventuresome tourist on a budget might stand in front of a roadside bar tipping back a cold beer alongside a group of workers who are also drinking a Red Stripe, as they relax after a long day of work. Even those who keep to the compounds of their resorts find their source of drink at poolside or at the hotel lounge during sunset "happy hour" or cocktail hour that is provided in their hotel package. No matter the location, catering to so many distinctive tastes and styles, the offer of drink and the promise of food are complemented by the sound of music in the air. At the end of the day, Negril sheds its laid-back ambience and becomes a hot spot for all kinds of entertainment for all ages.

Accordingly, the best of service is rendered by the silent hand of proficiency and face-to-face encounters are rewarded with customer satisfaction.

Figure 6.1   Sunset in Negril. A fiery ball sunset ends the day in Negril, and the night life begins. *Source:* Courtesy of Negril Chamber of Commerce.

Working in the food, beverage, and entertainment sector demands a required skill level. These skills are attained via training and acquired competence whereby a high quality of service is achieved through hard work and evolves toward a degree of professionalism in the industry.

HOSPITALITY 101

Human resource development is regarded as a critical aspect of hospitality education in Jamaica. The goal of the area of tourist development is to keep moving toward an era of professionalization, characterized by highly trained and educated employees and continuous improvement (Jayawardena and Crick 2000). One of the methods of training includes OJT—on-the-job training—which helps to maintain standards by having managers, supervisors, or fellow employees coach individuals in the most effective way to do the required job (Kum et al. 2005, 187). Those researchers also warn that while training may provide employees with the necessary skills, it does not ensure they will effectively utilize these skills and provide good service. Nonetheless, before training was institutionalized, for example, by UWI programs, HEART/Trust/National Training Agency, and the Jamaica Hotel and Tourist Association, staff was trained by experienced in-house employees in OJT style. The origin story of Sandals Resorts, recounted in *All That's Good* (Jaccarino 2005), reveals the importance and success of this kind of training. Training becomes a fundamental element that cuts across all areas of tourism, defining the quality of the personal interactions synonymous with good relations with the tourism staff. "The hotel business is a people business and you have to please your customers," remarked an associate in the Sandals Resorts (Jaccarino 2005). Thus, there is discussion about dining decorum, an essential element in all-inclusive tourism.

THE DINING ROOM

The quality of the food, the presentation of food, and the tasteful (according to the clientele) choices on a menu are critical in the all-inclusive resort business. One of the ways to guarantee a satisfied visitor was to make sure that three elements of food service are in place: adhering to the basics of dining room etiquette, how well the table is set, and finally how to serve guests. All who work in dining rooms were schooled in what is expected of them by their performance in achieving the ultimate goal of the returning visitor. Needless to say, those expectations often depended on the location of the dining space. Visitors who expect exceptional service might be satisfied in the formal settings that are distinctively addressed in the high-end resorts. Then, there are establishments that seek to emulate that upscale model and try to do their best. Those that feature a Jamaican homestyle atmosphere must also exceed in making customers satisfied, as do those in the corporate decorum of fast food. Thus, a major factor in the world of tourism and in the hospitality industry period is the quality and quantity of service provided to the paying customer through the presentation of food. Besides buffet-style dining, most of the waitstaff in Negril is charged with taking orders, recommending dishes based on guests' tastes and preferences (important in high-end resorts), delivering food to the tables, accepting payment (not necessary in all-inclusive resorts), but most importantly, ensuring an outstanding guest experience, even in the world of fast food.

These are the elements of service for eight women whose "workdays" get underway at the start of sundown. They are located in three different sites. The workplaces are in big all-inclusive (exclusive) resorts, and in food and beverage establishments with music and dancing that welcomes everyone. On all accounts, training and quality service were tantamount in securing customer satisfaction. The importance of a positive, enjoyable dining experience for the visitor is paramount here for these women workers. To begin, the chapter describes the experiences of a table captain, an activities director, a fashion show coordinator, and a floor show dancer. Each of the four women was employed in all-inclusive properties.

TABLE CAPTAIN

"Donna" began her employment in the hotel industry as a member of the wait staff at a Kingston hotel in the mid-1990s. She moved to Negril a few years later, following her husband, who as a musician (a side man) performed in all the big hotels on the north coast and, when possible, outside of Jamaica. When the job of a table captain opened up in the evening shift, Donna seized the opportunity. With this job, there are the incentives of extra money as well as increased responsibilities. Luckily enough, this came about at the time that Donna's husband was not on the road with his band and he could take care of and help with the homework of their two daughters, ages 13 and 7 years.

Among Donna's duties as the table captain is the extra responsibility of overseeing the transition of the main dining area from a late lunchtime poolside setting to a more formal one. There is sense of elegance in the dining area, as each table is set with a runner with coordinating centerpiece, cloth napkins, appropriate placemats, a full set of cutlery and glassware that includes water, and glasses for both white and red wine. Donna is considered a central figure in the waitstaff and is in charge of providing guests a gracious dining experience. Donna supervises the waitstaff and writes comments on their performances. She sees this position as one that has potential for advancement in her current place of work and in the larger world of the hotel industry. Still, she said, "The salary is small and the hours are long, but that's what hotel work is about."

In school, Donna did not "take her subjects seriously and [was] not focused at t'oll." Donna got pregnant at age 17, and her mother helped her with childcare. At that time, she secured a job as a banquet waiter, followed by taking a full-time job in the hotel's coffee shop. Donna remarked, "Working in tourism is an excellent way to make a living, especially in you have a service that is needed and make so people can manage it." She has "no problem" with the way visitors treat her as "some are more friendly than others." In the tourist business, Donna said, "There should more equal respect [in terms of] what an individual can or cannot do based on their capabilities." This Donna said was in reference to her duties as table captain when she must remind waitstaff of what is needed for the table, how to announce the specials not listed on the standard menu to the table, and other points that can be missed if the staff is not paying attention. One day, she was admonishing a young man when she recited the rules of being a member of the waitstaff , including the importance of good work ethics, such as looking good, standing tall, and listening to the requests of the guests. "You want them to give a good report to the manager so they come back." The reward of good service comes from her own trajectory from being a banquet waitstaff who just delivered the meal, whisked away dirty dishes, brushed off the crumbs on the table, and maybe refilled a guest's water glass. Now as table captain, she contemplates more than this job in her future with "the goal of taking correspondence courses in public relations."

BEVERLY AT THE VILLAGE

Each resort has its own agenda as to how to present a gratifying experience for their guests' trip to Negril. In a family-centered resort, there is an array of activities that are noted on the event board and printed on a weekly schedule. Copies of the activities schedule were available in the lobby, stacked near the cash register at the gift shop, and found neatly arranged near the dining area podium. This is the domain of the activities director ,whose job it is to make sure all guests are entertained by personal interaction—the managed heart.

Beverly is the activities director at the "Village," an all-inclusive near the center of Negril. A tall, athletically fit young woman, Bev was an elementary school teacher before coming to work at the Village. She approaches her work like a classroom. Events are scheduled and posted with certain kinds of tourists in mind. In the early part of the day, Bev organizes garden walks for those who are up and about. During these walks, which are just a turn of the grounds, Bev identifies the flowers and fauna around the resort. This activity does not generate a crowd, but it is something announced on the event board. By lunchtime, Bev has handed out table tennis paddles and made sure guests know where to return the equipment when they are finished with their games. Beach volleyball is arranged for 3 p.m. after the heat of the day dissipates. Bev not only encourage guests to join in on the games but also goes out of her way to recruit them off of their chaise lounges where they lay on the beach. Both player and cheerleader, Bev observes which of the guests would be good candidates for her evening event, the after-dinner entertainment and dancing.

As the sun slowly sinks into the sea, it is dinnertime, and as is the case for most all-inclusive resorts, guests are required to "dress up," meaning *no* swimwear, bare feet, or bare chests in the evening in the dining area. The kind of dress appropriate for many, like the Village, is classed "resort casual" with notions of "tasteful" clothing as an unwritten rule. Only the very exclusive resorts mandate that men wear jackets while dining in their specially reserved restaurants. Even so, at the Village the dining tables are set with at least six utensils, the appropriate number of glasses, and a change of plates between courses. The waitstaff is on high-performance mode.

Bev circulates around the room at about 8:45 p.m. as the dining is winding down and the band is tuning up. Gone is Bev's daytime attire of shorts and cross-trainers. She now wears a halter top and a long, brightly colored gauze skirt that a complemented by the dressy gold sandals on her feet. Bev approaches the guests she recognizes from the beach and invites herself to join them at their table. She is laughing, joking, and being an outgoing, friendly personality.

While the band is playing, Bev circles the dining room as she personally meets and greets as many guests as possible. Bev encourages all to get up on the dance floor. For families, the first dancers are usually children. Bev calls out to the children by name after checking with their parents or peers for the name if she has not memorized it yet. Soon the limbo pole is brought out and Bev implores the children to bring their parents with them to participate in the spectacle. By this time, the rum punch has dulled inhibitions of the adults, and the adventuresome ones are egged on by Bev to go low and lower under the pole. Prizes are awarded that usually consist of a bottle of rum or a coupon for an ice cream sundae to be redeemed at the snack bar the next day.

The dancing continues but the audience changes with families retiring to their rooms, towing their children behind them. In the meantime, the beat will go on for a few hours more, and those who took dance lessons, or refresher lessons, test their skills on the dance floor. Two days earlier, some of these guests learned to move with the reggae beat, thanks to the international influence of Bob Marley, and eventually figured out the fast-stepping jump-up *soca* from Trinidad and the Eastern Caribbean. In those lessons, Bev exposes guests to the differences among Caribbean cultures by recognizing the distinctive dance forms of the region. Bev dances with numerous partners and groups, encouraging them all to jump up!

After a while, the band begins to play popular U.S. music by Michael Jackson, Stevie Wonder, and others. Just before midnight, the band ends its session, and recorded music comes over the loudspeakers. Bev tells the guests gathered around her that it is time for her to leave the Village—after all, she is not on vacation, so she has to go to work in the morning—and they all laugh. She hopes that they had a good time, and that tomorrow there will be lots more of fun in store for them.

THE FASHION SHOW

In the early days of Air Jamaica (1968–2011), circa 1972, during mid-flights from New York to Montego Bay, a few of the flight attendants disappeared from the aisles, and quickly emerged wearing beautifully designed outfits. Another member of the crew became a fashion commentator while the captured audience of passengers were entertained by a fashion show in the air. Although Air Jamaica no longer exists, fashion shows are still a form of entertainment throughout Jamaica. Moreover, these shows are now held in the dining areas of resorts in Negril.

"SAND AND SEA" RESORT FASHION

Sand and Sea is a small all-inclusive resort near the end of the seven miles of the beach close to Negril Centre. The resort is a family business and also owns a few gift shops around town. Those shops stock all forms of souvenirs like bags of Blue Mountain coffee, Appleton Rum, spices such as Pick-a-Pepper sauce, towels that say "Jamaica," a large assortment of T-shirts, and other sundries. In addition, the shops also sell resort clothing such as swimwear, cover-ups, sarongs, sun dresses, halter tops, and Kariba or guayabera shirts for men. Although there is a gift shop on the premises of Sand and Sea, one of their family stores located in a shopping plaza down the road provides the clothing for the weekly after-dinner entertainment, the fashion show.

A collection of brightly colored island-inspired clothing is delivered to "Sand and Sea" for "Hazel" to select for the show. The clothing fabrics are of cotton or poly-mixed that are tie-dyed and usually printed with island motifs such as palm trees and flowers. Another task for the fashion show requires Hazel to recruit a model or two from the resorts' guests. Hazel usually approaches one or two guests to volunteer to model. Often without too much prodding, someone happily agrees to participate. Hazel's mainstay of models are members of the entertainment staff, such as "Luke," who officiates beach activities, Marguerite, the concierge, and perhaps another member of the activities staff who is available. For this show, she approaches two pre-teen guests to join in the fun and secures the permission of their parents for them to participate in the show. The night before, Hazel reminded her staff to bring appropriate footwear for the show, such as dressy sandals and perhaps high heels for the women and slip-on shoes for the men. On the day before, right after lunch, the young guests went to Hazel's office, tried on the clothes they were to model, and were given instructions as how to do a model's strut while navigating around the dining tables to display the fashions.

To accompany the fashion show, Hazel already selected recorded music of a smooth jazz steel drum band variety that was suitable for the models to walk around the room without tripping or colliding with tables. The music set the mood for the fashions on display, such as beach wear or casual resort wear for the evening. Hazel serves as the fashion commentator for the show and also models an outfit. She pushes the play button to start the show. In addition to describing the outfits and how beautiful they are, part of her commentary also announces the price of each outfit and where those items can be purchased—in the gift shop or the shop down the road. The models prance and twirl down and around the room, smiling, and waving to guests. Everyone claps and "oohs and ahhs" to specific performances. The pre-teen girl guests are thrilled to have their makeup applied and hair styled for the event, all done by Hazel's talented hands. The girls do a great job and are rewarded for their efforts by Hazel, the other staff models, and by the clapping and cheering audience. The girls are given T-shirts as a thank you for their participation. With great excitement, they both announce that they cannot wait to tell their friends back home that they learned how to strut like a model and wore lots of makeup. Their families are pleased with this activity, and all prove to be very satisfied visitors to Negril.

FLOOR SHOW

If one of the goals of tourism development is professionalization and a highly trained staff, then one can look no further but to the nightly entertainment of a floor show in one of the large all-inclusive hotels. The show takes place in a space that can also accommodate big events such as conventions, and other large occasions. During the height of the tourist season, the "Colonnade" room is configured into a theater format with chairs lined up in rows with aisles on each side. Guests take the first available seat as the room quickly fills up with vacationers, all in anticipation of the evening entertainment.

On the stage to the right is the band, which begins to play slow-paced reggae and calypso tunes to entertain the guests while they wait for the main event. The music has the crowd toe-tapping to the rhythm, while the spotlights came on and off setting the stage for things to come. The house lights dim and the master of ceremonies come onto the stage and welcomes the audience to Jamaica, to Negril, and most importantly to the resort. This evening's entertainment highlights one of Jamaica's greatest cultural assets: music and dance. The lights go down and the spotlights come up on a group of dancers dressed in "traditional" Jamaican peasant garb. By their grace and technique, these are professional dancers; they pirouette and sing along with the traditional Jamaican folk songs. Each set sees a few costume changes, as the MC provides commentary about island life, the meaning of the flag, and the motto of the nation, all set to the movements of the dancers, who appear in different vignettes. The audience is reminded of Harry Belafonte, with the dancers enacting the words of the songs he made popular, such as "The Banana Boat Song" and "Matilda." The dancers move with grace and style with their rendition of "Island in the Sun" and exhibit fun and excitement extenuated by drum rim shots. To close the show, a medley of Bob Marley hit songs is staged, with dancers performing reggae and dancehall moves while the audience dances in their seats. As the band ends their session, the guests exit the room, dancing down the aisles after they thoroughly enjoyed the evening.

"Joy," one of the dancers, was already dressed back into her street clothes when I approached to ask if she would mind doing a quick interview about her work as a dancer. Joy said she had just a few moments to spare while she took off her theatrical makeup. She explained that all of the dancers were leaving Negril shortly and were waiting on a van to drive them back to Kingston. Joy was a graduate of the Edna Manley College of the Visual and Performing Arts (EMCVPA). The BFA (bachelor fine arts) dance program at EMCVPA included courses in ballet, modern technique, folk technique, choreography, and dance history, among other subjects, including math, history, and modern languages. All of the dancers in this show are alumni of Edna Manley and had taken classes with one another. When the hotel was built, the word went out it was looking for dancers for the nightly entertainment. At first, Joy said she was hesitant because she did not want to be a spectacle of dancing under the limbo poll so "typical of north coast tourism." However, when she learned of the ideas of that year's (2014) showcase, which would be classy, culturally appropriate, and professionally choreographed, she signed a contract. Joy performed once a week at the resort from the beginning of January to mid-April and was paid a "decent" salary. Earlier that day, Joy and the group left Kingston midmorning and arrived in Negril four hours later. They changed into practice clothes, stretched, and warmed-up, and then went straight into rehearsal for the evening's performance. In that night's performance in Negril, Joy was able to use all of her dance techniques as cultural expressions to inform tourists of the richness of Jamaican culture. Joy and company did their jobs quite well and received an enthusiastic response by the audience.

In the all-inclusive resorts, there is entertainment. However, before a plate is served, there are cooks and chefs busy at work preparing the food. The next section addresses the history of ingredients of Jamaican food. Attention is paid to Jamaican food, tourist food, and fine dining, then centers on the interlocutors who were cooks and chefs in Negril.

SAVORY SAUCES AND MORE

Up and down the road, the aroma of food on the grill and on the industrial stove brings in a crowd and for some offers adventure in sensory delights. Negril can be the mecca to enjoy all things Jamaican, especially food. For the most part, all-inclusive resorts offer "typical" Jamaican food as a special night event, served buffet style with a steel drum band playing in the background. However, it is in the dining rooms of small hotels, in bars and restaurants, in small shops and Jerk chicken stands on Norman Manley Boulevard or on the West End that one finds the food for which Negril and Jamaica are famous. On the dinner menu are the staples of akee (ripe pods of the fruit are cooked with onions, tomatoes, and thyme), saltfish, callaloo (similar to spinach, cooked with tomatoes and sweet peppers), plantain, bread fruit or a yam, browned chicken, and beef stew, all seasoned with savory sauces, and spiced with scotch bonnet pepper, to mention just a few popular choices for visitors and Negrillians alike. Moreover, these traditional foodstuffs are homegrown and come out of the pots and stoves and experiences of the enslaved Africans who were brought to Jamaica during the slave trade.

In 1743, Captain Bligh, of Mutiny on the Bounty fame, brought akee and bread fruit tree seedlings as well as pineapples and perhaps bananas to the British colonies. Salted (salt used as a preservative) cod fish, a product of Nova Scotia, Canada, was added to this mix. The cultures of the enslaved from the Bight of Benin (West Africa) added their culinary knowledge of hot peppers, black pepper, yams, beans, to the indigenous pimento (allspice), which together all contributed to traditional Jamaican cuisine. This included preparing jerk meat and chicken seasoned with all of the spices just mentioned. Further, following the emancipation of African-descendent Jamaicans, the British colonial regime brought subcontinent Indians (1845) and Chinese (1854) to Jamaica as indentured labor to work on the colony's sugar plantations. The introduction of these two culinary traditions, infused with the African-based food, expanded the Jamaican palate to include curry, turmeric, roti, naan, fish sauce, bok choy, onion, jackfruit, and other food groups.

This section focuses on women who work in the culinary arena. One is a cook of traditional Jamaican food served in a bar/restaurant, while the other is chef who prepares French as well as Jamaican cuisines in the in-house dining in a beachside hotel. Each woman represents those who are able to seize opportunities in the culinary arts that require a high level of skills and training.

All along Norman Manley Blvd, there are numerous places to choose from for night-time dining. "Sandra" is the cook at "My Place," a well-known establishment for traditional Jamaican food and locally caught fish. This family owned restaurant has a barkeep who is also a family member and another relative as one of the waitstaff. Sandra became the cook following her mother, who helped to start this family business. At home, she learned to cook from her grandmother as well as her mother. However, she thinks she was like "an apprentice" when it came to learning to prepare food in quantity. This was OJT.

The night I interviewed her, the restaurant had ten tables, and all were full, and there was a line of people waiting outside of the door. It was August and Negril was very busy. During independence celebrations when Jamaicans living abroad come home for a holiday and Dream Week, a week-long party/summer music festival makes Negril a very busy place. Sandra works really long hours due to the uptick of visitors with little time for conversation, but she allowed me to observe her. In the morning all of the prep work was done, as cooking traditional style requires lots of cutting, peeling, soaking, and seasoning of the food. The menu did not vary except to state what fish is available on a given day. That night, "the fish of the day" was Escovitch red snapper served with a piece of "bammy," a fried cassava bread. Escovitch fish is very popular with the young crowd at the Dream Week festival. In advance, Sandra ordered an extra supply of take-away containers to accommodate the Dream Week crowd and asked her mother to help with the extra prep workload. The local fishermen have been familiar with Sandra's family for years and have a long-standing relationship by providing fish to "My Place." Midmorning, a fisherman came to the back door of the restaurant with his catch in a cooler. He caught the fish earlier in the day and Sandra was his primary customer. She bought his entire catch of snapper, most of which are of similar size.

All of the food is fresh, but to prepare an ample supply of the pickling sauce for the fish, Sandra started early slicing onion, scotch bonnet pepper, and some carrot, adding some to the marinade in a container with a bit of vinegar to season. The rest was set aside. The fish were seasoned and drenched with flour. When an order was placed, the fish was hard fried then doused with the sliced vegetables with a liberal amount of the vinegar mix to complete the dish.

After a busy day of preparing dishes on the menu, Sandra directed her cousin and her mother to clean up the kitchen for the day. Her cousin, the barkeep, stayed on for a few more hours and then closed the restaurant way after midnight.

Restaurants in Negril offer Italian, Mexican, Chinese, German, and Rastafarian vegan cuisine. American fast food such as pizza is served in many establishments, with a franchise of Burger King located in Negril Centre.

With so many places to choose from, the upscale restaurant in the Charela Inn stands out because it offers a blend of French cuisine with a touch of "Jamaican spice" and a generous selection of imported wines, served with exceptional care. Established in the mid-1980s by Daniel and the late Sylvie Grizzle, the Inn, a four-star hotel created an extraordinary dining experience for visitors. Moreover, Charela Inn is a perfect example of how a well-trained staff provides excellent service, yielding over forty years of repeat visitors as hotel guests as well as diners.

Daniel Grizzle owned a 174-acre farm in nearby Hanover. Thus, in 1983, when the Grizzles opened their restaurant, Sylvie, French by birth and Jamaican by marriage, was the head chef while Daniel functioned as both cashier and waiter (1994, 3). Before there was a concerted effort of the farm-to-table movement, the Grizzles' farm supplied fruits (in season), spices, vegetables, dairy products, poultry, sheep, cattle, and goats. They also sell their produce and meat to other restaurants in the area. Charela also serves its own baked goods, such as croissants for breakfast and rolls for evening dining. Keeping with the ethos of the farm-to-table movement, "a beautiful freshwater spring that emerges from the rock . . . void of chemicals or other water system impurities" (charlea.inn.com).

In one interview before her passing, "Mrs. G" stated, "Almost of the staff are women" and that she considered them "reliable and willing to succeed in the business and assume the responsibilities of the job . . . more loyal and try hard as a team." In the late 1990s, when she handed over much of the kitchen duties, "Mrs. G" hired a woman. She noted that the chef at Charela Inn was a woman who was "reliable because she was the primary wage earner for the family and perhaps for a whole 'tribe.'" Good training and keeping good staff was essential when companioned with the Charela operational concept of teamwork. Mrs. G remarked that "working as a team profited for everyone resulting with very little turn-over." An example of this can be found in the 2020 Negril Guide, which featured a Charela Inn chef who worked there for sixteen years and learned her skill and craft under the tutelage of "one of the best cooks in Negril," the late Sylvie Grizzle.

One of the goals that Mrs. G maintained was "to provide the perfect environment for guests to relax, get some peace, enjoy some aspects of our culture, and to dine on delicious Jamaican and French food." In this way, Charela Inn and Le Vendome, its upscale restaurant, have moved forward with professionalism and a highly trained and committed staff. Teamwork evokes a work ethos emanating from the staff, whose service renders more than a standard performance. It guarantees the return of a visitor looking for good food and relaxation.

WAIT, THERE IS MORE NIGHTLIFE

According to a tourist event calendar from 2019 (pre-pandemic), there is an abundance of entertainment in which to indulge in Negril. Not part of this study, but worth mentioning, is the Bob Marley Birthday Bash, which features five days of performances of top reggae stars, who come to Negril for this special event. Furthermore, the months of March and April bring Spring Breakers, whose sizeable numbers and money amplify the party atmosphere that is duplicated in August during Dream Weekend. In December, The Reggae Marathon course follows the coastline and Norman Manley Boulevard and allows runners to compete in the full marathon or half marathon or the 10 km depending on their ability. Accommodations for visitors rely on the range of lodging found in Negril and mentioned earlier in the book. Tourists can venture out of the comfort of the all-inclusive resort, depending on their budget and style of vacationing. Women who work in the entertainment subsector of tourism follow the seasons of the industry, including these special events that take place in Negril. Some of the best reggae bands in Jamaica, often from Kingston and often Bob Marley wannabes, show up in Negril.

Fun places are easy to find, as nearly everything is on Norman Manley Boulevard. "Unleash the animal" was the slogan for a well-known hot spot, which intimated the kind of party atmosphere found there, particularly during Spring Break and Dream Week. According to that online Negril calendar description,

> Red Stripe and rum punches are the drinks of choice for a crowd that really seems to enjoy their estrogen and testosterone highs. Fortunately, security at least appears tight, with a prominent sign in front that declares, NO PROSTITUTES OR GIGOLOS, NO DRUGS, NO SOLICITING, NO MISCONDUCT and a forbidding-looking bar that locks the place up tighter than a jail during off-hours. (visitjamaica.com: "Online hotel resource for travelers to Negril, Jamaica.")

Even with the expansion of tourism in Negril, which attracted the crowds to the Marley Bash and Dream Weekend, there was always the lure of sex in the tropics. Jamaica was associated not only (advertised as such) with sand, sea, and sex but also with a variety of practices (drugs, unbridled debauchery, and selling of sex) that were available for purchase. This is another kind of satisfied customer and the role that sex plays in tourism.

Throughout Jamaica's history of slavery, British colonizers exerted enormous control over Black women's bodies in different forms of work, including sexual labor. Significantly, the relations of power under slavery were not always monolithic or total. Female sexuality was a site for reconfigurations of power. In her landmark study, Kamala Kempadoo (1999, 7) noted that mulatto women sometimes made strategic use of their exoticized/eroticized status through self-employment of their sexual labor. Almost four hundred years later, women sex workers in Negril responded to consumer demand, now in its global form. As Beverly Mullings (2002, 57) remarked, "The growth of sex tourism in Jamaica reflects the power of increasingly globalized flows; of capital, policy directives, and information to draw places that were once 'off the map' into ever closer networks of commoditized trade and exchange." A 2010 Inter Press Service article featured a dancer named "Misty" who worked in an exotic adult nightclub in Negril. At age 24, Misty had been dancing with her skimpy costume for four years. At first, she was hesitant but found that she could "make good money" and could take care of her two-year-old daughter. Plus, the men tipped well. She remarked, "I hope to become the best dancer in here."

Negril was particularly impacted by sex tourism because of its popularity as a tourist destination sector and the increased impoverishment of the people of the country at large, especially young women. Although commercial sex is against the law, and in 2007 the government implemented the Trafficking in Persons Act, prostitution is tolerated because it is a means of survival. In 2016, there was a call from Negril stakeholders to regularize sex work, to decriminalize it, and make it safe for workers from a health point of view. Chairman of Couples Resort John Issa remarked, "I think you can have some sort of coexistence, but it would have to be confined, controlled, and managed, because we are not going to get rid of it." In "Tourism and Sexual Violence and Exploitation in Jamaica," Cruz et al. (2019) noted that the people they interviewed were not driven into sex work and were not prevented from exiting it by "human traffickers" but rather by economic need, and in the case of male and trans sex workers, by anti-gay prejudice.

One of the aims of this study was to represent as much as possible every type of job available to women. Because of its illegality, sex workers were the hardest women to make contact with for an interview. By chance, the conversations occurred with a handful of women tourist sex workers who were very young—ages 16 to 18. These young women saw this line of work as a way to have fun, bring a bit of foreign glamour into their lives, and to have access to their own money. All of them left school because of economic pressures from their families. There was just not enough money for school fees, lunch money, bus fare, uniforms, and the like for all of the children of the house. These young women saw tourists partying, dancing, and having fun in an international and U.S. style in which they wanted to partake. All of them were drawn to the trade when they saw their agemates that they knew at a local club pair off with foreign White men. They had heard stories of what was going on. They watched their acquaintances go to the supermarket and fill up a buggy with all kinds of foodstuffs that were well out of reach for their own families with the foreign White man paying for it all. They wanted to try this out for themselves.

One young woman who lost her virginity to a foreign male customer turned this personal sacrifice to her own advantage. This is just one narrative that tells how the global tourist market forces are at play in the sex trade. "Ronnie" said this about her work.

The first I decided to party with the tourists after I saw my friend do it. We got our clothes together so I would look good to them White men. Me and my friend went to the club that many of them came looking for a Jamaican girl. You cannot go to the hotel with them or go near the gates cause the security to run you off, even if you look good. They might try to talk to you, but you want a man who has plenty money, maybe for a week if you get lucky. My first try, I got my tight, tight jeans on, push up myself inna little top, put make up on and I gone. Go stand by the bar and order a drink and just watch for someone to stand by me. The room was a crush. A group of White guys comes up to me and my friend and we talk. They want to find ganja first then to hook up, but I didna want to waste my time on them. Then, my friend who did this before brings a guy over, kinda old, but not too old. We talk, he buys me a drink, we laugh and I look him over and he look at me. He wants me bad I can see. He was staying in a villa we can go there. I figure that since I was virgin, and I hear you can get more for it cause everyone says that you cannot get AIDS from a virgin. We get to the place. So we do it and he like it that I a real virgin and a young thing. At the end, I charge him three times more the price for the one time but he pays cause he says it was so fresh. I figure how can I make like I am a virgin all of the time?

For Ronnie, the lack of skills and education coupled with a heightened sense of consumerism provided the downside of the success of tourism in Negril. Sex work will remain as long as the lure of sex and drugs is part of Negril's appeal.

HOW STELLA WAS REWORKED

To this very day, tourist agents as well as hoteliers constantly receive requests from couples seeking romance, fun, and sun while looking for the location of where *How Stella Got Her Groove Back* was filmed. They want to replicate that glamorous and sexual experience. Then there are women who ask the same question: Where can they too get their groove on? Like the character Stella, they want to have a romantic vacation with a dark stranger who heightened their sense of adventure in their vacation experience.

For those who do engage in what is called "romance tourism," the nomenclature of "Stella" has meaning for both the Jamaican men who call out to female tourists regardless of age or race, and the women who engage in these affairs for a week or the duration of their stay. Stella wannabes who seek out men who are performing a "Rent-a-Dread" or other kinds of relationships are not categorized as in engaging in illicit sex, but it is still a transactional activity. Here Jamaican male prowess becomes a commodity with a temporary cash flow deemed as the outcome. The often-cited "Romance Tourism, Gender, Race and Power in Jamaica" (Pruitt and La Font 1995, 436) concluded,

> The Jamaican who assumes that all tourists are wealthy may be disillusioned when he discovers that the object of his attentions is spending money freely in order to have a special vacation but is neither rich nor extravagant once the holiday is over.

In 2019, the conjecture was that Negril's reputation of being a laid-back kind of place strengthened the temporality of those relations. Men were quite aware of conditions of asking for more, such as sponsorship to the United States. No matter what, even for Jamaican men who approach women tourists in anticipation of a week or so of access to food, gifts, and cash, there remain issues of control. It is the women tourists who have the ability, the power to call these actions a romantic interlude or sexual harassment. As the interlocutors who participated in Williams' (2018) research conducted in the mid-2000s pronounced, the power was securely in their hands and the Jamaican men whom they encountered were more provocateurs than anything. According to Bianca Williams (2018, 33), in her ethnography *The Pursuit of Happiness*, some of the African American women in the study were targets of sexual harassment. On the beach, in town, and in nightclubs they frequented, they were called out as "Stella" by those assuming that these African American women were looking for a romantic tryst with a Jamaican man while on vacation, just like in the movie *How Stella Got Her*

Groove Back. Williams remarked that the name calling at times was annoying to the women who were actually in pursuit of happiness, peace, and valuable emotional ties to assuage the racism and sexism they faced back in the United States.

AFTER MIDNIGHT

Arlie Hochschild (1983, 5) remarked that "emotional labor style of the offering (service) that makes this effort, the emotional sense of intent and purpose that it becomes part of the service." As each woman's workday was illustrated, the tenor and level of personal engagement showed how these actions became part of her own performance—expending that emotional labor that educed how advanced training evolved into professionalism that led to high-quality service that is ultimately rewarded. The combining of the mind and feeling that is conveyed to others showed up in face-to-face encounters of teaching pre-teen girls to strut like a model or encouraging guests to dance in order to understand the culture in which they are vacationing. OTJ was the main operative in all of these cases save for the graduates of Edna Manley, whose professionalism was engrained in their study but then applied in their stage performances. As for Ronnie, the teenage sex worker, the lure of fast money was fueled by a heightened sense of consumerism that she saw displayed by her peers who were tourists. Her experience might be closer to that of Misty, whose story the IPS reported.

Conclusion

Women Tourist Workers in the Capital of Casual

There are so many panoramic images of Negril's white sandy beach that it would be remiss not to include one in this conclusion (figure C.1). In this photo, the white rim that meets various shades of maritime blue is the acclaimed beach that has a worldwide reputation as a tourist destination. The river, various kinds of housing dispersed between the green trees and fauna, and the road are clear from above. Although you cannot see them from this bird's eye high angle, there are women who are hard at work in a variety not jobs in those buildings, on the road, and on the beach. *Women and Tourist Work in Jamaica* provided a perspective of the world's fastest-growing economic sector that is women centered, reliant on women's labor, and made successful in Negril by women's managerial and creative expertise. In their capable hands, women tourist workers carve out a space making a living for themselves and their families.

There were a number of moving parts in the telling of this story, framed by feminist thinking, recounting a brief history of Jamaica's tourist sector and its many reincarnations. Most importantly, there is an unfolding of the evolution of a remote fishing village that used mixed-wage employment and women's multiple occupational cultivation and marketing as sources of livelihoods into becoming "the Capital of Casual." As colonialism was on the wane, post–World War II economic planning of the era engineered use of the latest innovative technologies of transportation and communication, with the growth of the middle class to promote tourism as an economic strategy. Negril's late entry into Jamaica's tourist planning was facilitated by the already-tested partnering of local capitalist elites with government policies and agency bodies like the Urban Development Corporation (UDC). Government funds were secured by international lending agencies such as the World Bank alongside of grants and loans from the United States and other nations. Borrowing has

Figure C.1   Satellite View of Negril. *Source:* © Vonkara1/iStock/Getty Images Plus/Getty Images.

led to Jamaica's national debt soaring at different moments in time, but it is partly a responsible driver of Jamaica's tourist sector.

WORKFORCE LOCATION

Beneficial to Jamaica's tourism was the historical importance of women's labor force participation, listed in 2020 at 58.5 percent, according to World Bank data derived from the International Labour Organization. As the worldwide cheerleader for tourism, the UNWTO promoted the cause of women, who as the catalysts of tourism provide essential sources of low-wage labor as well as a way for women's empowerment. *Women and Tourist Work in Jamaica* showed how the expansion of neoliberalism that garnered capital through private and public investment relied on and is rewarded by a predominately female workforce whose service, hospitality, and professionalism produce this "product" called "tourism."

Jamaican sociologist Orlando Patterson (2019) wrote that Jamaica was blessed with near-perfect tourism resources—sand, sea, and sun. Further, despite all the calamities facing the country, Jamaica has a large, growing, and highly competitive tourist industry. Compared with other Caribbean countries, Patterson said, "the industry retains a much higher percentage of its earnings (70 percent vs. 30 percent for the region) and it has the highest percent of local ownership and management" (335). It is not just the ownership and management, but the workforce, specifically, a female workforce. One of the reasons for the success of Jamaica's tourist industry is the quality and quantity of essential labor of women who make all of this economic magic happen. Service, hospitality, and professionalism encompass this magic, as represented in *Women and Tourist Work in Jamaica.*

ANTHROPOLOGICAL PRACTICES AND THOUGHTS

The framework of this study was guided by anthropological methods and theories. Snowball sampling was the mainstay of the day that allowed for cordial personal contacts for conversation, often continued over the course of two decades of this study through a variety of forms of communication. Copious notes were taken, guided at first by an interview, then just referred to during informal settings as conversations took place on the fly. Archival, online searches with UN agencies, hospitality gurus, and tourist guide publications provided data and information concerning the industry. Key in the theory building of tourism studies was Erve Chambers, whose "face-to-face encounters" became almost a code of what was occurring between women workers and visitors. George Gmelch's use of narratives in his study on tourism in Barbados was a valuable resource. Also located in Barbados was Carla Freeman's examination of neoliberalism, gender, and women's lives in the up-and-coming middle class that helped to center notions of female entrepreneurship. Other studies carried out by anthropologists in the Caribbean illustrated the value of using tourism as a way of interrogating the preservation of national identities such as found in the British Virgin Islands (BVI) examined by Collen B. Cohen. Then there was the valuable work by Carla Guerron Montero in two Caribbean locations. The intrusion of tourism on the small island of Carricou saw return migrants called Kayaks forging a way to maintain the integrity of the "Big Drum" traditional festival so it could remain an important cultural element through means of celebrating its authenticity on home turf, while staging that celebration elsewhere on the island as a tourist attraction. Guerron Montero's other work focused on Boco del Turo, an island off the coast of Panama. For Afro-Antillean communities composed of descendants of nineteenth-century workers, Montero illustrated the ways of using their own heritage (an amalgamation of Jamaican, Barbadian, and other West Indian cultures) as a fusion West Indian style for tourist consumption. By engaging directly with global trends and the imaginary of the Caribbean Afro-Antillean culture, they elevated their place in Panamanian national identity formation.

From the early days of the 1950s through this current project, anthropologists carried out research and lived in Negril for various durations. Beginning with William Davenport, followed twenty years later by Deborah D'Amico-Samuels and Barbara Olson and two decades later by Barry Chevannes (ISER project), anthropologists have captured the essence of social change as it was happening in Negril. First there was the promise of economic fulfillment followed by the realization of the unevenness of tourism in this community. Chevannes's collection of life stories appearing in the Organization of American States (OAS) and the U.S.-funded UWI Institute of Social and Economic Research documented recounted family members' self-examinations of their working-class migrant cashflow in anticipation of a boom in tourism. Sociologists Hopeton Dunn and Leith Dunn's 2002 survey and community evaluations addressed the rise of tensions between the actions of poor women and men vendors hustling and hassling tourists, which was deemed a negative in the tourist industry, requiring Jamaican government policing of the issue.

Besides the social scientists, *Women and Tourist Work in Jamaica* relied on literature, poetry, and essays that helped to frame the intentions of Caribbean lives over time. Black feminist poet June Jordan's essay helped to locate this researcher's own positionality in terms of reciprocity and respect with the women with whom she worked.

THE INTERLOCUTORS

Representing the ongoing process of tourism in Negril are the women interlocutors who shared their work stories, their experiences in the sector, and the impact they feel that tourism has on their making a living. Included in their narratives are cultural practices of Jamaicaness and skin-teeth that become part and parcel of providing good service in a hospitable climate for visitors to Negril. Further, the highly rated success of Negril complements the cultural attribute of reciprocity and friendliness embedded in Jamaicaness. Women tourist workers who extend their valuable comportment of Jamaicaness and receive reciprocal responses are essentially at the top order of rendering service. On the other hand, one of the verbal cultural practices that helped Jamaicans, particularly for women who manage the rude behavior of tourists, was to show "skin-teeth." Used with passive aggressive verbal or body language, skin-teeth demonstrated how inequitable relations are resolved in a tourist context where "the customer is always right."

For the past fifty years, women tourist workers have made their livelihoods in Negril and essential providers in this workforce. In the course of this research, thirty women interlocutors were shadowed in their workday and during the night. Their own agency in this business of personal service was evoked in their Jamaicaness, their use of emotional labor, their savvy commercial sense, and the dissonance they conveyed via cultural practices of showing skin-teeth.

Without knowing so, *Women and Tourist Work in Jamaica* followed a template set in place for understanding tourism in the Caribbean by Graham Dann (1988). He noted that tourism was a "product" and a resort was conceived in terms of its pull factors—for example, sunshine, friendly people, and beaches—and advertised by means of brochures, travel logs, films, and other promotional devices. Tourism has become a "new kind of sugar," a monocrop of inculcating power relationships based on racial, class, and gender considerations carried forward to the present. Elitism, sexism, lack of equal educational opportunity, unfilled aspirations, and seasonal employment pervade the industry. *Women and Tourist Work in Jamaica* filled in the blanks of who is doing the work in tourism, and not by just listing labor force statistics. As a matter of fact, *Women and Tourist Work* provided a close-up and direct look at how tourist women workers' own agency is brought to bear as they manage the face-to-face encounters, those that are affable and those centered around inequitable social relations. Dissent is not tolerated, especially if you need or want to keep your job. Therefore, a sense of social harmony keeps guests pleased with their vacation experience and planning to return once again to Negril.

ETHNOGRAPHIC SENSIBILITIES

What is described in *Women and Tourist Work in Jamaica* was based on ethnographic sensibilities that honor mutual respect, engaged in responsible research, and provided an opportunity for exchange with interlocutors that promotes increased accuracy in writing and publishing. Pseudonyms were used (except for three women) for the people and places, which followed a chronology demarcating change over time. Observations and sometimes participation in events occurred on resort beachfront properties or in garden gazebos, where they were viewed and commented on in the planning of assorted social activities by hotel social event coordinators and underwater when snorkeling. In attempts to recruit wait staff and cooks to participate in the study, meals were eaten at a few restaurants where a bar tab was maintained. Craft vendors were observed as they made raffia embellishments on straw baskets. The waitstaff was surveyed as they set up tables for breakfast then cleared the food debris when they set up the place settings for lunch, followed once again by clearing the dirty dishes and food remains. After all that, tables were rearranged and table décor for dinner was put in place. All of this human activity was documented in handwritten notes and typed on a dysfunctional small computer that proved to be useless. There were no audio recordings. The result was a study on tourism without centering on tourists.

The only of interlocutor name used in *Women and Tourist Work in Jamaica* was Mrs. Sophie Grizzle-Roumel, while Jackie Lewis's narrative relied on a published interview. The locations of work sites were also given basic descriptive names except when indicated, such as Jackie's on the Reef. The travel-ready intersectional method of gender, color/race, age, and class allowed for categorizing the sources of livelihoods and the social meanings of difference that played out in those work spaces. In addition, shared in the narratives were social factors such as age, educational attainment, and length of employments. Among the number of moving parts were various kinds of service, quality of service, accommodations, face-to-face encounters, hospitality, and professionalism, to name but a few. As chapter 3—"Women, Work, and Tourism"—showed, there were distinctive groupings by work sites: small business/entrepreneurs, hotel managers and professionals, housekeepers, food and beverage workers, entertainment, and local artists and crafts. The women who participated in this study ranged in age from sixteen to seventies, two were White women who married Jamaican men, only one was a native Negrillian and their educational levels ranged from third-grade schooling to postgraduate study.

Throughout the ethnographic chapters, as a group, all thirty women saw working in tourism as a good source of making a living. Even the disgruntled few did not malign the sector itself, but just groused about the lack of opportunities afforded to them by their lack of skill/education. Often those complaints centered around tourists who tipped poorly or not at all and indeed had racial overtones, and then there were the security guards who hassled them on the beach. After a few years, when their kiosks literally fell apart, at the behest of the Negril Chamber of Commerce (NCC), the kiosk hair braiders became associated with a hotel and became part of the hotel retinue. Hair braiders and massage workers were relocated to beachfront spots equipped with umbrellas, chairs, canopy tents, and massage tables. As illustrated by the housekeepers in this study, who basically had six years of education they saw tourism as their source of their survival. All of these women maximized their employment earnings by extending their Jamaicaness through good service and were handsomely rewarded by visitors. A handful of young professionals looked forward to advancing their careers in the sector or in a related field such as public relations.

NEGRIL CHAMBER OF COMMERCE

All of the interlocutors who owned hotels and cottages and other small businesses were members of the NCC. Their establishments benefited from their NCC membership through lobbying efforts for road maintenance, improved water and electricity service, internet access, waste management, and promotion efforts beyond what the Jamaican Tourist Board offered.

There are many business opportunities for women in tourism in Negril. Besides owning or managing family-owned lodgings or seeking career advancement in the industry, women are hard at work in "paradise." The strong aspects of the gender system do appear in Negril. However, they are mitigated by the tenacity of women interlocutors allowing ingenuity, on the job and advanced training, and the positive aspects of emotional labor to surface. The narratives of the women tourist workers in the subsector of food, beverage, and entertainment illustrated this point. The best of service was rendered by the silent hand of proficiency and face-to-face encounters that were rewarded with customer satisfaction. High-caliber service was acquired via training, and high-quality work evolved toward professionalism in the industry.

Long before the pandemic, the NCC promoted its communities with the help of the *Guide to Negril*. Significant on the NCC website was a donation solicitation button for NCC community efforts that has recently included calls for recovering efforts for COVID-19, which put most of the working population out of work. Part of the mission of the *Guide* is to immerse visitors into the life of Negril by educating them and making them a member of the community. This strategy maintained that when you come back, you can see where your dollars, Euros, and other currencies were well spent. In this way a visitor can relish in their own brand of Jamaicaness that marks the ambiance of Negril and its laid-back but embracing aura.

SPANISH INVASION

For years Sandals Resorts and other locally owned and operated hotels dominated Jamaica. But for the last ten years, as Patterson (2019) and other scholars noted, more Spanish-operated all-inclusive resorts have been coming on stream than ever before. Besides RIU, there is Azul, a part of the Iberostar Corporation. As of fall, 2021, Royalton, another Spanish-owned corporation, will merge with Marriott International. As feared and voiced by "Sybil" in chapter 4, RIU and other conglomerates were so heavily invested in tourism in Negril to the point that the village was experiencing a full-fledged invasion. Furthermore, it was reported that the ethos of Negril, the laid-back and highly touted friendly interactions, were also being undermined by these international chains, who demand corporate modes of deportment by staff rather than the personal face-to-face encounters associated with the Capital of Casual.

A pre-pandemic 2019/2020 Sandals Resorts campaign seen in print and on television drew the viewer into the realm of high expectations of a high-end vacation experience in one of its luxurious hotels including those located in Jamaica and in Negril (cnn.com: October 7, 2019). Standing out against a background of crystal, clear water, swinging in a suspended woven seat for two, there is an attentive Black butler who carries a tray of drinks to the beautiful people who await his service. This advertisement was not about the "typical" tourist package, but one that portrays an extravagant Jamaican tourist experience that can be bought for an upscale price.

The image does two things. First, the most luxurious accommodations require the critical presence of a personal butler, a Black male servant whose subservience has a long history and connotes prestige afforded by the most sophisticated visitors. Second, there were invisible hands behind the personal butler where a phalanx of women provided all of the support for those opulent trappings, such as daily changing the linen, putting a piece of chocolate on the pillow, and reminding the butler to reserve the guest's golf tee time or spa appointment. Although butler service was not the norm, the phalanx of women workers who perform the essential elements on a high level is the gold standard in tourism and is found throughout the sector in small and medium establishments. Just how successful Jamaican women tourist workers can be is found in the numbers of repeat visitors who come back to Jamaica again and again. The Jamaican Tourist Board keeps tabs on the number of repeat visitors that is confirmed by data and word of mouth. Tourism in Negril relies on the many roles that women tourist workers play for its image, service, and accountability. It is on that very local level of interaction and business transactions that businesswomen make their contributions to the success of Jamaica's tourist industry and make a living for themselves and for their families and for some to make a profit.

THE COVID-19 PANDEMIC

In December 2019, Jamaica's tourist industry was poised to exceed its expectations of the upcoming season. Nine months later Jamaica had 6,408 confirmed cases, 1,624 recovered, and 101 deaths due to COVID-19. Jamaicans were ordered to wear masks with the understanding of the risks of this public health crisis. Jamaicans recognized the peril to their lives following other public health crises. However, dealing with illness due to mosquitos was one thing; COVID-19 was another.

In Negril, the pandemic hit hard because of the number of international visitors who brought the virus, and the lack of tourists whose dollars and euros fueled the economy of the community. Not only was COVID disruptive of business, but also the resumption of business to normal levels was unknown. What was clear that because business/tourism was shut down, the livelihoods of the workers also came to a standstill. One early marker of the pandemic occurred when a boutique hotel on the cliffs saw a dramatic number of cancellations mostly coming from its European clientele. Further, sixty-four out of its seventy-six employees were sent home, presumably for the duration of the hotel closure (Durrant Pate, Jamaica Observer West, March 20, 2020). The article stated that several other hotels in Negril were shuttered due to a lack of business, while the few that remained open have significantly reduced their workforce. RIU Negril, which employed over three hundred workers, announced that in early March 2020, the hotel was closed. There was optimism voiced by the leaders in Montego Bay's tourism that the industry would "ride out the storm." However, those who rely on work in the sector were set adrift without any economic safety net.

Through aggressive public health actions, the Ministry of Health, Immigration and Customs, and other agencies took control as best as possible to slow the spread of the virus in the general population. In the meantime, in Negril, like the rest of the "tourist corridor," tourist workers who were released from their workplaces or, because of the lack of customers, had little money for food for themselves and their families. Hotelier and philanthropist Paul Salmon used his talents and his long-term relationships with the people of Negril to establish a campaign to feed the needy of this tourist-dependent community. Using a donate button on the Rockhouse Foundation webpage, Salmon and his team solicited funds to feed desperate families. Targeting the families already in partnership in their education program network, they made weekly distributions of food staples to families in and around Negril. A safety net program aimed at delivering benefits to the most needy and vulnerable in the society steadily expanded to include additional families struggling to meet basic nutritional needs. The Rockhouse Foundation was making weekly food distributions and sustaining approximately a thousand people in the community with food staples. RIU reopened in April 2021, celebrating twenty years doing business in Negril. Even this four-hundred-room property followed COVID-19 Jamaican Government protocols, and it is assumed that the staff was vaccinated as suggested by the NCC.

LAST WORD

Writing in the aftermath of a destructive hurricane season, Jamaican economist Michael Witter (2017) remarked that hurricanes Harvey and Irma demonstrated the vulnerability of both rich and poor Caribbean societies and the extreme fragility of the poor islanders. Other Caribbean countries that were not as damaged saw greed and survival intertwined, and the precariousness of life in Dominica and Puerto Rico after [Hurricane] Maria will take some time to be fully understood. He goes on to say that Caribbean tourism (in reference to Jamaica) that focuses on imported consumption goods offers abundant opportunities for a concerted effort to bring the private sector into the conversation to implement sustainable development. But as Witter warily notes, "Big corporations seek to get as much as they can for as little as they can give up" (Witter 2017, 5). In sum, the urgency of sustainable socioeconomic development is the only strategy for survival, as seen in efforts by the environmental sensibilities of the NCC, the businesses that understand that climate change is real, and members of the Negril community who see the erosion of the beach and the destruction of the reef as all warning signs. On top of making a living, women and their families know that tourism has both strengths and weaknesses.

Women and Tourist Work in Jamaica illustrated how women tourist workers in their own voices express how they felt about making a living in personal service to others, how lack of education stymies opportunity, how face-to-face encounters draw on cultural practices that assuage the social inequities that surface and are assured by reciprocal kindness sometimes housed in managed hearts. High-caliber service and hospitality require emotional acts and can be monetarily rewarded. Over the years of various Jamaican Tourist Board slogans, women tourist workers showed that Jamaica is more than a beach. It is also a country, with one love. So, once you go, you will know that Jamaica is a home of the "alright," setting the tempo of the heartbeat of the world.

References

Altexsoft. 2019, December 19. "A Brief History Lesson on Travel: Why, How, and Where We Traveled in the 1920s." https://www.altexsoft.com/blog/travel-in-the-1920s

Althoff, Susanne. 2020. *Breaking the System of Women Entrepreneurs.* Boston: Beacon Press.

Amber, Jeannie. 2005, September. "The Wind at My Back by Jackie Lewis." *Essence Magazine.*

Anderson, M. J. 2020, September 20. "On the Rights of Women." *Boston Globe,* N 8–9.

Anderson, Patricia Y. 1986. "Conclusion: Women in the Caribbean: Afterview." *Social and Economic Studies* 35:2:291–324.

Babb, Florence. 2010. "Tourism in Formerly Off-Limits in Latin America." *Anthropology News* 51:8:21.

———. 2011. *The Tourism Encounter.* Stanford: Stanford University Press.

Baker, Christopher. 2000. *Jamaica.* Victoria, Australia: Lonely Planet Publishers.

Banks, Ingrid. 2000. *Hair Matters: Beauty, Power, and Black Women's Consciousness.* New York: NYU Press.

Barriteau, Eudine. 1998. "Theorizing Gender Systems and the Project of Modernity in the Twentieth-Century Caribbean." *Feminist Review* 59:186–210.

———. 2002. "Women Entrepreneurs and Economic Marginality." In *Gendered Realities*, edited by Patricia Mohammed, 221–48. Barbados, Jamaica, and Trinidad and Tobago: University of the West Indies Press.

Barrow, Christine. 1996. *Family in the Caribbean: Themes and Perspectives.* Kingston, Jamaica: Ian Randle Publishers.

Beckwith, Martha W. 1928. "Notes on Jamaican Ethnobotany." In *Memoirs of the American Folklore Society*, p. xxi. New York: The American Folklore Society.

Bolles, A. Lynn. 1992. "Sand, Sea and the Forbidden: Media Images of Race and Gender in Jamaican Tourism." *Transforming Anthropology* 3:1:30–35.

———. 1996. *Sister Jamaica*. Lanham, MD: University Press of America.

———. 1997. "Women as Category of Analysis in Scholarship on Tourism: Jamaican Women and Tourism Employment." In *Tourism and Culture*, edited by Erve Chambers, 72–92. Albany, NY: SUNY Press.

———. 2001. "Flying the Love Bird and Other Tourist Jobs in Jamaica: Women Tourist Workers in Jamaica." In *Sister Circle: Black Women and Work*, edited by Sharon Harley and the Black Women and Work Collective, 29–47. New Brunswick, NJ: Rutgers University Press.

———. 2002. "Women Workers and Global Tourism in Jamaica." In *Black Women, Globalization, and Economic Justice*, edited by Filomena C. Steady, 228–44. Rochester, VT: Schenkman Books.

———. 2005. "The Golden Goose: Race, Gender and Global Tourism in Jamaica." In *Cultura y Desarrollo* (special issue, Culture and Diversity and Tourism) 4:82–94.

Bolles, A. Lynn. 2015 "Academics and Praxis: Caribbean Feminisms." In *Transatlantic Feminisms*, edited by Cheryl R. Rodriguez, Dzodi Tasikata, and Akosua Adomako Ampofo. pp. 63–77. Lanham, MD: Lexington Books.

Braithwaite, Edward Kamau. 1973. *The Arrivants: A New-World Trilogy*. Oxford: Oxford University Press.

Brennen, Denise. 2004. *What's Love Got to Do with It?* Durham, NC: Duke University Press.

Browne, Katherine. 2004. *Creole Economics: Caribbean Cunning under the French Flag*. Austin: University of Texas Press.

Bruner, Edward. 2005. *Culture on Tour*. Chicago: University Chicago Press.

———. 2012. "Around the World in Sixty Years." In *The Restless Anthropologist*, edited by Alma Gottlieb, 138–59. Chicago: University of Chicago Press.

Bush, Barbara. 1990. *Slave Women in Caribbean Society*. Bloomington: Indiana University Press.

Caribbean hotel and tourism.com. 2019. Oxford Economics Travel and Tourism. TE Oxford Travel Tourism in Jamaica.

Carnegie, Charles. 1986. "Inter-Island Trading in the Eastern Caribbean and the Informal Sector Concept." In *Margins, Caribbean Environment*, edited by Yves Renard, 123–50. Santo Domingo: ENDA-Carib.

Carter, Sara, and Tom Cannon. 1992. *Women as Entrepreneurs*. London: Academic Press.

Cassidy, F. G., and R. B. LePage. 1990. *Dictionary of Jamaican English*. Cambridge: Cambridge University Press.

Casson, Lionel. 1994. *Travel in the Ancient World*. Baltimore: Johns Hopkins University Press.

Chambers, Erve. 1997. *Native Tours*. Long Grove, IL: Waveland Press.

———. 2000. *Native Tours: The Anthropology of Travel and Tourism*, 2nd ed. Prospect Heights, IL: Waveland Press.

Comitas, Lambros. 1963. "Occupational Multiplicity in Rural Jamaica." In *Proceedings of the American Ethnological Society*, edited by V. Garfield and E. Friedl, 41–50. Seattle: University of Washington Press.

Communicationarts.com/project/7851/Jamaica. (accessed September 9, 2020).

Conde, Maryse. 1992. *The Tree of Life: A Novel of the Caribbean*. New York: Ballentine Books.

Conklin, Mark. 2000. *Banana Shout*. Irving, TX: Fusion Press.

Crenshaw, Kimberlee. 1989. "Demarginalizing the Intersection of Race and Sex: A Black Feminist Critique of Antidiscrimination Doctrine, Feminist Theory, and Antiracist Politics." *University of Chicago Legal Forum*, volume 40, 139–67.

Crick, Anne P. 2001. "Personalized Service in the New Economy." *Journal of Eastern Caribbean Studies* 26:1:1–20.

Cruz, K., J. O'Connell Davidson, and J. Sanchez Taylor. 2019. "Tourism and Sexual Violence and Exploitation in Jamaica: Contesting the 'Trafficking and Modern Slavery' Frame." *Journal of British Academy* 7:1:191–216.

D'Amico-Samuels, Deborah. 1986. *You Can Get Me Out of the Race*. Ph.D. dissertation, City University of New York.

Dann, Graham M. S. 1988. "Tourism Research on the Caribbean: An Evaluation." *Leisure Science* 10:261–80.

Davenport, William. 1956. *A Comparative Study of Two Jamaican Fishing Communities*. Ph.D. dissertation, Yale University.

Davis, Dana A., and Christa Craven. 2016. *Feminist Ethnography*. Lanham, MD: Rowman and Littlefield.

Davis, Omar, and Patricia Y. Anderson. 1987. *The Impact of the Recession and Adjustment Policies on Poor Urban Women in Jamaica*. Report prepared for UNICEF.

Deere, Carmen Diana, et al. 1990. *In the Shadows of the Sun*. Boulder, CO: Westview Press.

DiGiovine, Michael. 2010, November. "UNESCO's Vehicles for Peace." *Anthropology News* 51:8:8.

Doyle, Dane, and Bernbach. www.ddb.com.

Dunn, Hopeton S., and Leith L. Dunn. 2002a. *People and Tourism*. Kingston, Jamaica: Arawak Publications.

———. 2002b. "Tourism and Popular Perceptions: Mapping Jamaican Attitudes." *Social and Economic Studies* 51:1:25–45.

Edwards, A. 2008, April 18. "Spanish Hotel Operations Loot Jamaica." *Jamaica Observer*.

Enloe, Cynthia. 1990. *Beaches, Bananas and Bases*. Berkeley: University of California Press.

Fernandez, Nadine 1999. "Back to the Future." In *Sun, Sex and Gold*, edited by Kamala Kempadoo, 81–89. Lanham, MD: Rowman and Littlefield.

———. 2010. *Revolutionizing Romance*. New Brunswick, NJ: Rutgers University Press.

Figueroa, Esther, dir. 2006. *Jamaica for Sale*. Vagabond Media. Kingston, JA: Jamaica Environment Trust.

Freeman, Carla. 2014. *Entrepreneurial Selves: Neoliberalism, Respectability and the Making of a Caribbean Middle Class*. Durham, NC: Duke University Press.

Friel, Bob. 2000, June/July. "Adventures in All-Inclusive Land" *Caribbean Travel and Life* 15:4.

Ghodsee, Kristen. 2005. *The Red Riviera*. Durham, NC: Duke University Press.

Gibson, Heather. 2001. "Gender in Tourism: Theoretical Perspectives." In *Women as Producers and Consumers of Tourism in Developing Regions*, edited by Y. Apostolopoulos, S. Sonmez, and D. Timothy, 19–43. Westport, CT: Praeger.

The Gleaner. 2016, March 28. "Call for Order—Negril Stakeholders."

———. 2021, January 20. "Largest Decline."

———. 2021, March 2. "Growth and Jobs."

Gmelch, George. 2012. *Behind the Smile: The Working Lives of Caribbean Tourism*, 2nd ed. Bloomington: University of Indiana Press.

Gmelch, Sharon B. 2004. *Tourists and Tourism: A Reader*. Prospect Heights, IL: Waveland Press.

Gordon, Derek. 1989. "Women, Work and Social Mobility in Post-War Jamaica" In *Women and the Sexual Division of Labor in the Caribbean*, ed. Keith Hart, 67–80. Kingston, Jamaica: The Consortium Graduate School of Social Science.

Graburn, Nelson, and Roland Moore. 1994. "Anthropological Research on Tourism." In *Travel, Tourism and Hospitality Research*, ed. J. R. Brent Richies and C. R. Goeldner. New York: John Wiley and Sons.

Green, Cecilia. 1999. "Afro-Caribbean Women Workers and the Conundrum of Respectability: A Historic Analysis." Unpublished Presentation at the American Anthropological Association Meetings, Chicago, IL.

Guerrón Montero, Carla. 2015. "Tourism, Cultural Heritage and Regional Identities in the Isle of Spice." *Journal of Tourism and Cultural Change* 13:1:1–21.

———. 2020. *From Temporary Migrants to Permanent Attractions*. Tuscaloosa: University of Alabama Press.

Hall, Catherine 1995. "Gender Politics and Imperial Politics: Rethinking the Histories of Empire." In *Engendering History: Caribbean Women in Historical Perspective*, edited by Verne Shepard, Bridget Brereton, and Barbara Bailey, eds. Kingston, Jamaica: Ian Randle.

Hall, Stuart, 1997. *Cultural Representation and Signifying Practices*. Thousand Oaks, CA:. Sage Press.

Harrison, Faye V. 1997. "The Gendered Politics and Violence of Structural Adjustment." In *Situated Lives: Gender and Culture in Everyday Life*, edited by L. Lamphere, H. Ragone, and P. Zavella, 451–68. New York: Routledge.

Hayle, Carolyn. 2005. "Tourism in Jamaica: The Impact of the Past on the future." In *Caribbean Tourism: Visions, Missions and Challenges*, edited by Chandana Jayawardena, 119–38. Kingston, Jamaica: Ian Randle Publishers.

Henderson, James. 2013, March 16. "Two Sides of Negril." *The Telegraph*.

Higman, B. W. 1979. *Slave Population and Economy in Jamaica 1807–1834*. Cambridge: Cambridge University Press.

Hochschild, Arlie R. 1983. *The Managed Heart*. Berkeley: University of California Press.

http://www.makingcents.com.

http://www/ofad/prg/gender.

Hurston, Zora Neale. 1981 [1938]. *Tell My Horse*. Berkeley, CA: Turtle Island.

"India-Popular Women's Micro-Enterprises in Manipur."

Institute of Social and Economic Research, Organization of American States. 1991. *Research Project on the Social and Economic Impact of Tourism in Jamaica.* Unpublished final report.

Inter Press Service. 2010, March. "Jamaica: Other Side of Paradise." www.ipsnews .net/2010/03/jamaica-the-other-side-of paradise.

Issa, John, and Chandana Jaywardena. 2005. "All Inclusive Business in the Caribbean." In *Caribbean Tourism: People, Service and Hospitality*, edited by Chandana Jaywardena, 223–35. Kingston, Jamaica: Ian Randle Publishers.

Jaccarino, Pamela Lerner. 2005. *All That's Good.* Kingston, Jamaica: Sandow Media.

Jamaica Hotel and Tourist Association (JHTA). 2012. "Our Jamaica." http://ssvu .com/hcmedia/docs/Jamaicameetingplanerspaced (accessed 3/25/21).

Jamaican Tourist Board. jtbonline.org (accessed October 26, 2019).

James, Daniel. *Layer Culture* (blog). www.layerculture.com.

www.jamaicans.com/jamaicatouristboard (accessed 9/4/2020).

The Jamaica Gleaner. 2019, March 16. "Environmental Protection Trust Restores Negril Coral Reefs."

Jamaica Information Service. jis.gov.jm/information/Jamaican history (accessed 9/11/2020).

———. statinja.gov.jm/LabourForce.

Jamaica Observer West.

Jamaica Observer. 2019, September 23. "Bartlett Names Tourism Pension Scheme."

Jaywardena, Chandana. 2002. "Future Challenges for Tourism in the Caribbean." *Social and Economic Studies* 51:1:1–23.

Johnson, Renee. May 20, 2020 "TPDCo urges compliance with Safety Protocols."

Jordan, June. 1985. "Report from the Bahamas." In *Feminist Theory Reader*, 2nd ed., edited by Carol McCann and Seung-kyung Kim, 160–68. New York: Routledge.

Joseph, Naline R. 2005. "Service vs. Servility: Creating a Service Attitude in Caribbean Hospitality Industry." In *Caribbean Tourism: People, Service, and Hospitality*, edited by Chandana Jayawardena, 171–81. Kingston, Jamaica: Ian Randle Publishers.

Kempadoo, Kamala. 1999. "Continuities and Change." In *Sun, Sex, and Gold*, edited by Kamala Kempadoo, 3–36. Lanham, MD: Rowman and Littlefield.

Kincaid, Jamaica. 1988. *A Small Place.* New York: Penguin.

Kinnard, Vivian, Uma Kothari, and Derek Hall. 1994. "Tourism: Gender Perspectives." In *Tourism: A Gender Analysis*, edited by Vivian Kinnard and Derek Hall. Chichester: Wiley; pp. 210–216.

Klein, Laura F. 2004. *Women and Men in World Cultures.* Boston: McGraw Hill.

Knight, Franlklin. 1978. *The Caribbean.* New York: Oxford University Press.

Knoema. N.d. "Jamaica: Tourism." http://kKnoema.com/atlas/Jamaica/topics/tourism (accessed 10/8/2021).

Koulma-Castelberg, Mary. 1991. "Greek Women and Tourism: Women's Cooperatives as an Alternative Form of Organization." In *Working Women*, edited by N. Redflift and M. T. Sinclair, 197–212. New York: Routledge.

Kum, Jerry A., Roxanne Benjamin Hoppie, Rhona Crawford, and Oscar Pacheo Lopez. 2005. "An Insight on Training and Development in the Hospitality Sector in Jamaica." In *Caribbean Tourism: People, Service, and Hospitality*, edited by Chandana Jayawardena, 182–211. Kingston, Jamaica: Ian Randle Publishers.

Learch, Patricia, and Dian E. Levy. 1990. "A Solid Foundation: Predicting Success in Barbados's Tourist Industry." *Human Organization* 49:4:355–63.

LeFranc, Elsie. 1989. "Petty Trading and Labour Mobility: Higglers in the Kingston Metropolitan Area." In *Women and the Sexual Division of Labour in the Caribbean*, edited by Keith Hart, 99–132. Kingston, Jamaica: The Consortium Graduate School of Social Sciences, University of the West Indies, Mona.

Leite, Naomi M., and Nelson Graburn. 2009. "Anthropological Interventions in Tourism Studies." In *The SAGE Handbook of Tourism Studies*. London: SAGE.

Leite, Naomi M., Quetzil E. Castaneda, and Kathleen M. Adams. 2019. *The Ethnography of Tourism: Edward Bruner and Beyond*. Lanham, MD: Rowman and Littlefield.

Lofgren, Orvar. 2004. "The Global Beach." In *Tourists and Tourism*, edited by Sharon Gmelch, 35–53 Prospect Heights, IL: Waveland Press.

Mathurin Mair, Lucille. 1974. *A Historical Study of Women in Jamaica, 1655-1844*. Ph.D. dissertation, University of the West Indies.

———. 1975. *The Rebel Woman*. Kingston, Jamaica: African-Caribbean Institute.

Meade, Natalie. 2020, July 23. "After Beating Back the Coronavirus, Jamaica Prioritizes Tourism over Public Health." https://www.newyorker.com/news/news-desk/after-beating-back-the-coronavirus-jamaica-prioritizes-tourism-over-public-health.

Meeks, Brian. 2000. *Narratives of Resistance: Barbados, Jamaica and Trinidad and Tobago*. Kingston, Jamaica: University of the West Indies Press.

Mintz, Sidney W. 1974a. "The Historical Sociology of Jamaican Villages." In *Caribbean Transformations*, 157–79. Baltimore: The Johns Hopkins University Press.

———. 1974b. "The Origins of the Jamaican Market System." In *Caribbean Transformations*, 180–213. Baltimore: The Johns Hopkins University Press.

Mullings, Beverly. 2002. ""Globalization, Tourism and the International Sex Trade." In *Black Women, Globalization and Economic Justice*, edited by Filomena C. Steady, 295–329. Rochester, VT: Schenkman Books.

National Library of Jamaica. N.d. "The Slave Trade." https://nlj.gov.jm/slave-trade (accessed 10/8/2021).

Negril Chamber of Commerce. 2020. "Little Piece of Paradise." In *Negril Guide 2020*. Negril, Jamaica: Negril Chamber of Commerce.

Negril.com.

Nichols, Grace. 1992. "Skin-Teeth." In *Daughters of Africa*, edited by M. Busby, 797. New York: Ballantine Books.

Nixon, Angelique V. 2015. *Resisting Paradise*. Jackson. University of Mississippi Press.

Numbeo. N.d. "Cost of Living in Negril." www.numbeo.com/cost-of-living/in/Negril-Jamaica (accessed 10/8/2021).

Olsen, Barbara. 1997. "Environmentally Sustainable Development and Tourism: Lessons from Negril, Jamaica." *Human Organization* 56:3:285–93.

Outlook Magazine. 1994, July 17. "Daniel and Sylvie Grizzle: Making Strides in Tourism Together."

Patterson, Orlando. 2019. *The Confounding Island*. Cambridge, MA: Harvard University Press.

Patullo, Polly. 1996. *Last Resorts*. Kingston, Jamaica: Ian Randle Publishers.

Pike, Joe. 2009, December 19. "Super Clubs." *Travel Agent Central*.

Planning Institute of Jamaica. *2001 Economic and Social Survey of Jamaica*. Kingston, Jamaica: Planning Institute of Jamaica.

Poon, A. 1988. "Innovation and the Future of Caribbean Tourism." *Tourism Management*, Vol. 9, No. 3, 213–220.

Pratt, Mary Louise. 1992. *Imperial Eye: Travel Writing and Transculturation*. New York: Routledge.

Pritchard, Annett, Nigel Morgan, Irena Ateljevis, and Candice Harris, eds. 2007. *Tourism and Gender*. Oxfordshire, UK: CABI.

Pruitt, Deborah, and Suzanne La Font. 2004. "Romance Tourism, Gender, Race and Power." In *Tourists and Tourism*, edited by Sharon B. Gmelch, 317–35. Long Grove, IL: Waveland Press.

Quora.com. "Hospitality when you give selflessly." (accessed 3/2021)

Rhiney, Kevon C. 2012. "The Negril Tourism Industry: Growth, Challenges and Future Prospects." *Caribbean Journal of Earth Science* 43:25–34.

Robinson, Edward, and Malcom Hendry. 2012. "Coastal Change and Evolution at Negril." *Caribbean Journal of Earth Science* 43:3–9.

Roland, Kaifa. 2011. *Cuban Color in Tourism and La Lucha*. New York: Oxford University Press.

Rosenberg, Leah. 2010, February. "Tourism and the Birth of Jamaican Literature." *Jamaica Journal* 32:3:46–51.

Said, Edward. 2003. *Orientalism*, 25th anniversary edition. New York: Vintage.

Salazar, Noel B. 2005. "Tourism and Glocalization: Local Tour Guiding." *Annals of Tourism Research* 32:3:628–46.

———. 2011. "The Power of Imagination in Transnational Mobilities." *Identities* 18:576–98.

Scase, Richard, and Robert Goffee. 1980. *The Real World of the Small Business Owner*. London: Croom Helm Ltd.

Seaga, Edward. 2006. "Keynote Address: Tourism as the Driver of Change." In *Tourism as the Driver of Jamaican Economy*, edited by Kenneth O. Hall and Rheima Holding, xix–xxxix. Kingston, Jamaica: Ian Randle.

Sheller, Mimi. 2003. *Consuming the Caribbean*. London: Routledge

Shepard, Verene. 1999. *Women in Caribbean in History*. Kingston, Jamaica: Ian Randle Publishers.

Smith, Valene, ed. 1989. *Hosts and Guests*. Philadelphia: University Pennsylvania Press.

South Florida Caribbean News. 2020, April 1. "Jamaica Tourist Board Turns 65 Years Old Today." https://sflcn.com/jamaica-tourist-board-turns-65-years-old-today.

Statista.com/stastics/803809/Jamaica-gender-gap-labor-market/category (accessed 4/11/21).

Statistical Institute of Jamaica. statin.gov.jm (accessed October 29, 2019).

Statista.com/stastics/803809/Jamaica-gender-gap-labor-market/ category (accessed 4/11/21)

Steady, Filomina C. 2002. "Introduction." In *Black Women, Globalization, and Economic Justice*, edited by Filomina C. Steady, 1–37. Rochester, VT: Schenkman Books.

Stupart, Copeland, and Robert Shipley. 2013. "Jamaica's Tourism: Sun, Sea and Sand to Cultural Heritage." *Journal of Tourism Insights* 3:1:4. https://scholarworks.gvsu .edu/jti/vol3/iss1/4.

Swain, Margaret. 1995. "Gender and Tourism." *Annals of Tourism Research* 22:2:247–66.

Taylor, Frank F. 1993. *To Hell with Paradise: A History of the Jamaican Tourist Industry*. Pittsburgh, PA: University of Pittsburgh Press.

Thomas, Deborah A. 2004. *Modern Blackness*. Durham, NC: Duke University Press.

Thompson, Krista. 2006. *An Eye for the Tropics: Photography, Tourism and Framing the Caribbean Picturesque*. Durham, NC: Duke University Press.

Timothy, Dallen J. 2001. "Gender Relations in Tourism." In *Women as Producers and Consumers of Tourism in Developing Regions*, edited by Y. Postolopaulos, S. Sonmez, and D. Timothy, 235–48. Westport, CT: Praeger.

Tourism Product Development Company. www.tpdc.org/visitor-safety.

Tourism Review News (accessed 11/11/19).

Tourismanalytics.com/Caribbean202.

Travelagent.com/Caribbean. March 30, 2021.

Travelpluse.com (accessed October 29, 2019).

Tradingecomomics.com/Jamaicangovernment-debtstogdp.

Trines, Stefan. 2019, September 10. "Education in Jamaica." *World Education News and Reviews*. https://wenr.wes.org/2019/09/education-in-jamaica.

Turner, Matt. 2019. "James Bond Heads Back to Jamaica." *Travel Agent Central* (blog).

Ulysses, Gina A. 2007. *Downtown Ladies*. Chicago: University of Chicago Press.

United Nations Development Program. *Human Development Report*. New York and Oxford: Oxford University Press.

UNWTO. 2019, May. *World Tourism Barometer* 17: 2. https://www.e-unwto.org/toc/ wtobarometereng/17/2.

UNWTO database (accessed 4/15/2021).

UNWTO.org. history.

Urry, John. 1990. *The Tourist Gaze*. London: SAGE.

U.S. Central Intelligence Agency. 2021, September 23. "The World FactBook— Jamaica." https://www.cia.gov/the-world-factbook/countries/jamaica.

www.cdc.gov.

www.onecaribbean.org.

www.visitjamaica.com.

www.visitnegril.com.

Williams, Bianca C. 2018. *The Pursuit of Happiness*. Durham, NC: Duke University Press.

Winfrey, Oprah. 1997. "Special Message." In *Andre Talks Hair*. Saddle River, NJ: Prentice-Hall.

Witter, Michael. 2017. "Focus on Survival." *The Jamaica Gleaner*. http://jamaica -gleaner.com/article/focus/2017/001 (accessed 4/14/2020).

World Tourism Organization. 2019, December 3. "Global Report on Women in Tourism, Second Edition." https://www.unwto.org/publication/global-report -women-tourism-2-edition

World Tourism Organization. 2021, March 8. "News Release."

World Tourism Organization http://www.world-tourism.org/facts/trends/economy .htm.

World Tourism Organization. www.world-tourism.org.

Index

147

About the Author

A. Lynn Bolles is professor emerita of the Harriet Tubman Department of Women's, Gender, and Sexuality Studies and former affiliate faculty in anthropology, African American studies, comparative literature, and American studies at the University of Maryland College Park. Author of five books and over fifty book chapters and articles, she received numerous research grants from the University of Maryland, the Ford Foundation, and the Rockefeller Foundation, and was an NSF-funded ADVANCE professor. She was elected to the executive board of the American Anthropological Association and is past president of the Association of Black Anthropologists, Association of Feminist Anthropologists, and the Society of the Anthropology of North America. From 1992 to 1997, Bolles was on the executive council of the Caribbean Studies Association and served as CSA president from 1997 to 1998. Dr. Bolles was named Minority Faculty Member of the Year in 1992 at the University of Maryland College Park. In acknowledgment to the profession, she was given the 2013 Association of Black Anthropologists Legacy award. In 2014, Bolles was named UMD graduate school mentor of the year. Since retiring, Dr. Bolles has served as chair of assessment committees for the University of the West Indies. At present she is a member of a three-year Mellon Foundation grant team focusing on the pedagogy and student interactions among the African and African American Diaspora in Ghana and the United States. In November, Lynn Bolles will receive the American Anthropological Association "Gender Equity Award." Dr. Bolles earned a BA from Syracuse University with a double major in anthropology and English. At Rutgers (New Brunswick) she earned a master's in anthropology and in five years was awarded a PhD in social and cultural anthropology.

www.ingramcontent.com/pod-product-compliance
Lightning Source LLC
Chambersburg PA
CBHW050610280326
41932CB00016B/2986